Risk Management in Construction Projects

Edited by Nthatisi Khatleli

Published in London, United Kingdom

IntechOpen

Supporting open minds since 2005

Risk Management in Construction Projects
http://dx.doi.org/10.5772/intechopen.80628
Edited by Nthatisi Khatleli

Contributors
Yueting Hu, Janez Kušar, Tena Žužek, Tomaž Berlec, Lidija Rihar, Patrick Brockett, Linda Golden, John Betak, Ecem Tezel, Pinar Cakmak, Sarah Phoya, Krystyna Pietrzyk

First published in London, United Kingdom, 2019 by IntechOpen
IntechOpen is the global imprint of INTECHOPEN LIMITED, registered in England and Wales, registration number: 11086078, The Shard, 25th floor, 32 London Bridge Street
London, SE19SG - United Kingdom
Printed in Croatia

British Library Cataloguing-in-Publication Data
A catalogue record for this book is available from the British Library

Additional hard and PDF copies can be obtained from orders@intechopen.com

Risk Management in Construction Projects
Edited by Nthatisi Khatleli
p. cm.
Print ISBN 978-1-83962-216-8
Online ISBN 978-1-83962-217-5
eBook (PDF) ISBN 978-1-83962-218-2

Meet the editor

Dr. Nthatisi Khatleli is a senior lecturer at the University of the Witwatersrand in South Africa. He has a background in quantity surveying and he is still very much involved in that space. His research interest includes risk management in projects. He has also published extensively on public private partnerships, mega projects, and sustainability in construction.

Contents

Preface XI

Chapter 1 1
Standard Risk Management Model for Infrastructure Projects
by Lidija Rihar, Tena Žužek, Tomaž Berlec and Janez Kušar

Chapter 2 17
A New Mode of HSE Risk Management for Construction Projects
by Yueting Hu

Chapter 3 33
Different Market Methods for Transferring Financial Risks in Construction
by Patrick L. Brockett, Linda L. Golden and John Betak

Chapter 4 55
Holistic View on Multi-Stakeholders' Influence on Health and Safety Risk Management in Construction Projects in Tanzania
by Sarah Phoya and Krystyna Pietrzyk

Chapter 5 71
A Guide for Risk Management in Construction Projects: Present Knowledge and Future Directions
by Pinar Irlayici Cakmak and Ecem Tezel

Preface

Writing a book is a tedious undertaking but for an academic it is cathartic because you get an opportunity to tell the world about things that you have always wanted to. It gives ample time and space to explain your perspective without being too cautious of the space limitations. I have worked in the industry for more than 20 years as a quantity surveyor and I have always been interested in knowing more about what is risky, how to determine risk, and how to avoid it or minimize it but using systematic relevant approaches. I have come to realize that when it comes to risk you have to know the context and apply the tools that could be efficacious in that environment especially when the challenges are not purely technical.

When I took the challenge of editing this book I soon realized that one has to absorb a lot of information from many sectors and immediately guide the writer to stay very close and be relevant to the theme, but at the same time give them space to express themselves. It was not an easy task because a lot had to be chopped off that was not considered to be relevant to the theme. To those authors who did not make the cut I would love to encourage them to look for other opportunities to tell the world their perspectival appreciation of their chosen subject. Their contributions matter and their importance surpasses and goes beyond this particular publication. So not being published should not be regarded as a train smash.

Reading chapters was the most fulfilling part of the whole exercise because you get to learn many different approaches to risk. You also get to learn whether people have particular and/or different interpretations of risk. What I would love to have seen more of was how culture influences the way practitioners deal and appreciate risk. But what we have in this book is exceptional and intense in the sense that it equips the learner and gives an intimate panoramic view of how people interpret risk. Beyond just reading the chapters the administration work involved was handled superbly well and I would love to take this opportunity to thank the publishers for their professionalism, dedication, patience, and prowess in assisting me in making sure that only the work of the highest quality was published.

I thank everyone who has contributed to making this project a success. It is a shared glory and it should not belong to any one individual alone. Doing so will deprive it of its proper glow and dampen its deserved shine.

Dr. Nthatisi Khatleli
University of the Witwatersrand,
Johannesburg, South Africa

Chapter 1

Standard Risk Management Model for Infrastructure Projects

Lidija Rihar, Tena Žužek, Tomaž Berlec and Janez Kušar

Abstract

This paper outlines a risk management method that is based on the use of a standard risk management model and is adapted to the specific nature of infrastructure projects. The standard model can be used to identify and quantify unexpected events in planning and executing a project. The use of a risk map will also be illustrated. A risk map can serve to classify the identified and quantified risk events, depending on the expected loss, to critical risks that call for a more in-depth treatment, and non-critical risks that are normally not monitored, while no measures are foreseen in advance. A risk map is used to determine what the anticipated effects of the measures to mitigate the critical risks will be, and how the anticipated measures enable the transition from a critical risk to a non-critical risk. In this article, the suggested risk management is illustrated using the example of the erection of a reservoir for a hydroelectric power plant. The use of the proposed tools for the identification, assessment, prioritisation, and management of risks proved highly successful. With the use of the proposed risk model, the critical risk events were lowered under the acceptable level of the expected losses.

Keywords: risk management, standard model, risk map, risk control, hydroelectric power plant

1. Introduction

Infrastructure projects are one constant in our lives that interfere in our living environment and commonly involve huge investment costs. When managing such projects, the focus is mainly on the management of the content of work, times, resources, and costs. Risk management, however, is often neglected. Most frequently, the most important risks of the entire project are identified, and the measures to mitigate their consequences are prepared. Yet a project team lacks the time and motivation to prepare a more profound assessment of risks of individual components associated with the project.

The paper will illustrate the use of the standard risk management model, which includes the identification of risk event drivers, the assessment of probability of a risk event, and the identification of impact drivers caused by a risk event and the probability of its impact. The identified probability of the occurrence of a risk event and the probability of its impact serve as a basis for calculating the expected loss, most often in terms of time, money, or quality. The calculated losses can be represented in a so-called risk map, into which losses are plotted on the x-axis, while the product of the risk likelihood occurrence and its impact is plotted on the y-axis. A threshold line of expected losses divides the risks into critical risks (positioned above the threshold line of anticipated losses in the risk map) and less critical risks (positioned below the threshold line).

The standard model also allows an analysis of the consequences of the measures adopted and designed to eliminate or at least mitigate the expected risks both on the side of risk event drivers and on the side of risk impact drivers (there may be only a single or several drivers in both cases). In the risk map, the adopted measures represent a change in the risk position, the final goal being a shift of all critical risks below the threshold line of expected losses, i.e., below the limit of a still acceptable loss, by using the adopted measures in both risk factors.

The use of the suggested model will be illustrated using the example of an infrastructure project for the erection of a reservoir for a hydroelectric power plant on the Lower Sava River. The advantages and drawbacks of using the standard risk management model in the practical implementation of infrastructural (constructional) projects will be presented.

2. Review of references

Infrastructure projects most frequently involve the arrangement of an infrastructure building into a space (environment), which is why their success depends not only on internal factors, such as the client and contractor, but to a large extent also on external factors related to the environment. These factors aim to influence a project from various points of view; some of them support the project and want to make a positive contribution to the progress and success of the project, while others are completely or only partially against the execution of the project and are prepared to have a negative impact on the project. The execution of such projects is frequently considerably influenced by decisions of the government and the competent ministries. The mentioned impact factors may cause risk events on an infrastructure project, which may in turn have a very negative impact on the progress of the project, particularly on the execution time, on the costs, and often on the quality of the project deliverables.

Generally, risk management is a constituent part of the risk management strategy of a company and represents an important element in decision-making processes [1]. Infrastructure projects are particularly sensitive in terms of risks, because the risk events from similar previously executed projects only seldom repeat in a similar form and with a similar probability of their occurrence and consequences. Risk management in these projects is especially demanding, so it is important which risk management techniques are employed. Analyses show that financial and economic factors and quality are the most important risk factors that industry tries to avoid or transfer to other stakeholders [2].

The awareness or understanding that a risk may exist is in practice the most important aspect of risk analysis and management. How the participants understand the need for the treatment of each risk separately is therefore important for risk management [1].

As indicated by Hameed and Woo, numerous papers deal with the topic of risk management, yet the majority of research only includes risk management results from developed countries and only a very few from underdeveloped ones [3].

Slovenia belongs to the group of medium-developed European countries, even though a detailed analysis on the management of infrastructure projects has not yet been made.

Yafai [4] says that risks are treated in each infrastructure project differently, particularly based on an assessment of the probability of risk event occurrence and its impact and based on an individual project activity.

A variety of methodologies dealing with project management and consequently with the related risks can be found in literature. In practice, the most frequently applied methodology is the one proposed by the Project Management Institute [5]. Of the nine bodies of knowledge required for successful project management, it provides guidelines for risk identification, analysis, and response to project risks. Among other risk management methodologies, certain approaches warrant mention: PRINCE [6], which is mostly used in IT projects; DOD Risk Management [7], which is used for military industry projects; and a host of other methodologies [8]. A comparison of various risk management methodologies is shown in **Table 1** [9].

An important earmark in risk management is a proactive approach, which is explained in detail by Smith and Merritt in the book *Proactive Risk Management* [10]. They suggest various risk analysis and evaluation models (standard, simple, cascade, and Ishikawa risk models) and tools that project stakeholders can use for recording, prioritising, solving and monitoring reactions to project risks.

One important tool for project risk identification and analysis is a risk breakdown structure that systematically breaks down potential risks on several levels [11] and provides possible breakdowns for various project types. He suggests that the risk in infrastructure projects is divided into three levels: on the first level, he differentiates among risks that result from (1) environment, (2) contractors, (3) client, and (4) project.

Of course, major risk drivers may differ depending on the project type and the environment in which a project is carried out. Importantly, a project team responsible for project execution must identify all possible risk drivers on the project in question and break them down on several levels in order to facilitate a correlation between risk factors and project activities. An Ishikawa diagram can be used to identify risks [9].

PMBOK 2013	PRINCE2	DOD risk management
Plan risk management	Identify risk	Identify risk
Identify risks		
Perform quantitative risk analysis	Assess risk	Cluster analysis
Perform qualitative risk analysis		Risk mitigation planning
Plan risk responses	Risk plan	Risk mitigation plan implementation
Monitor and control risks	Implement and communicate	Risk tracking

Table 1.
Overview of several models for project risk management [9].

3. Risk management and infrastructure project

As infrastructure projects are most frequently integrated into a human living environment, risk factors (generators) appear in the risk assessment of such projects. In normal investment projects, these practically have no or in a few cases have a very small impact on project execution. Among the important impact factors that may cause risk events in infrastructure projects, the following warrant mention:

- Impact of space management institutions (Government of the Republic of Slovenia, ministry, local communities).
- Complicated procedures involving the integration of infrastructure buildings into a space.
- Local population, interest associations, environmental, and other organisations.
- In cases of public procurement, the possibility of appeal, auditing, and legal proceedings.
- Client's incapacity to finance the investment.
- Problems relating to solvency or even of contractor bankruptcies.

In investment projects, and even more particularly in infrastructure projects, additional risks can arise due to the following reasons:

- Poorly prepared plans for project execution without the use of adequate methods and techniques, usually also without a risk management plan.
- Poor and irregular reporting on work progress and actual costs.
- No reaction to deviations in the actual situation of the project from the plan.
- Frequent conflicts between parties executing the project because the responsibilities are not precisely defined.
- Execution time and cost pressures with a relatively low profit margin.
- Poor work safety due to pressures to produce good returns.

Experience in the management of investment projects has shown that, in a project planning phase, very frequently only the scope of a project is defined and a time and project cost plan are prepared. Normally, project managers do not deal with risks in the project planning phase. To assist project managers in risk management, a general model of project risk management was developed [9, 12], which originates from a particularly critical evaluation of the most frequently used project management procedures and especially project risks. The proposed model, which will be subsequently employed to manage a selected infrastructure project, is carried out in four phases and seven steps. Methods that a project team can use for efficient work are indicated for the execution of each individual step.

Figure 1 shows an amended project risk analysis model, namely, with reference to [9, 12], and a risk map is added for a qualitative and quantitative analysis of activity risks, for the classification of risks to critical and noncritical ones (step 4) and for planning of measures for risk management (step 5).

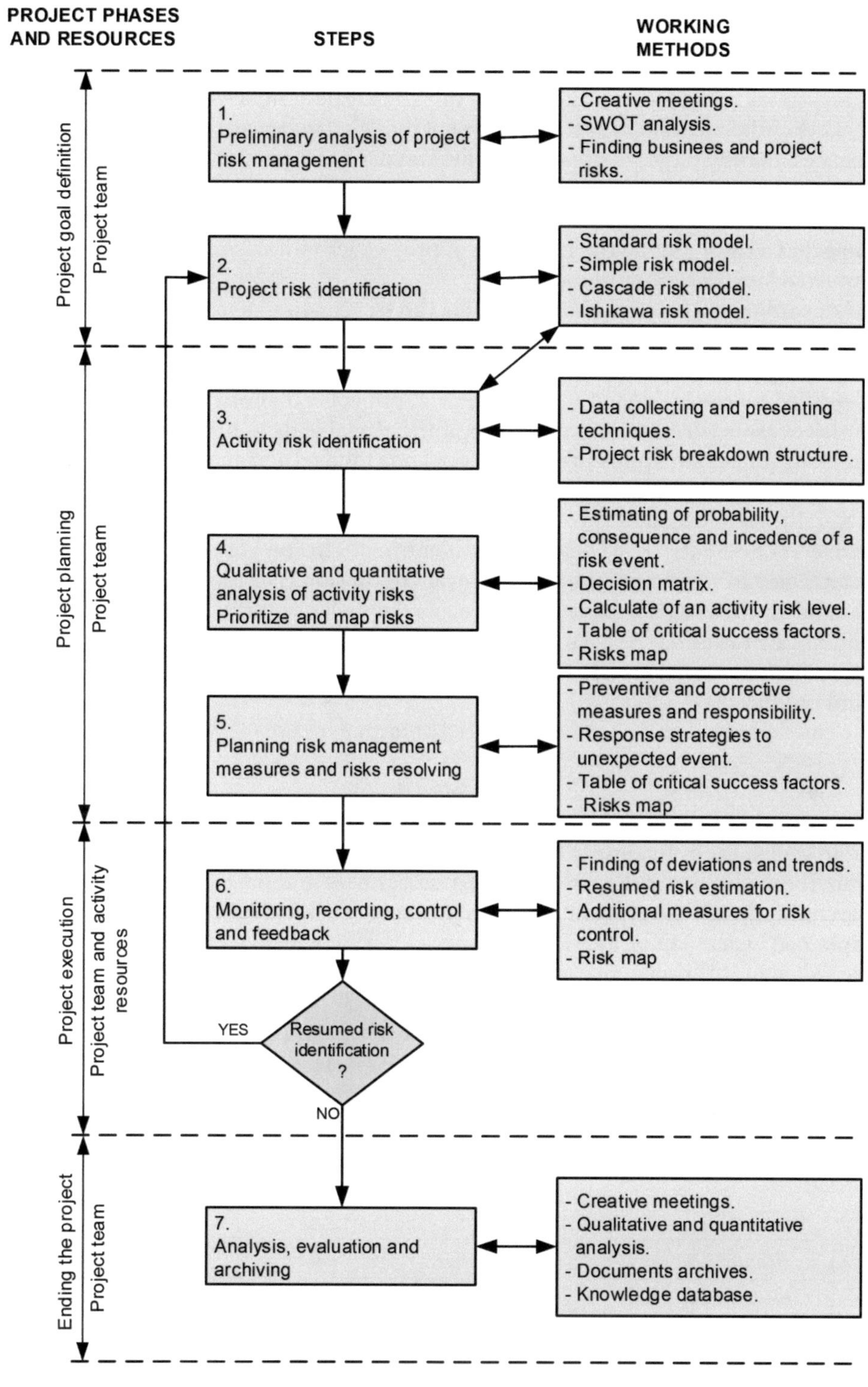

Figure 1.
An extended model of infrastructure project risk management.

As evident from **Figure 1**, the risks related to the entire project are first identified in steps 1 and 2. Various approaches can be used. Smith and Merrit [10] propose four different models for the identification and quantification of risks: standard, simple, cascade, and the Ishikawa models. Each of the proposed models has its advantages and disadvantages. When addressing the risks of an entire project, we concentrate on general questions, such as What is the risk, and what kind of loss can be expected if project execution is delayed by 6 months?

The proposed models for risk identification and quantification can also be used in steps 3 and 4, where individual risks can already be assigned to activities.

One of the mentioned risk management models can be used for further risk analysis. In this study the standard model was used to manage the activity risks in infrastructure projects. The reason for this decision lies in the fact that the model is simple to understand that it first identifies potential risk events and only then the impact of a risk event on the execution of project activities using a calculation of the expected loss (in time or money).

According to [10], the standard model can be visualised as shown in **Figure 2**.

In the standard model, a risk event is first identified. We can start from a previously prepared WBS/RBS matrix. One or several risk factors (drivers) can be identified for the incidence of a risk event. A project team must assess a probability of risk event occurrence *Pe* on the basis of the available data, on experience from previous similar situations or by using methods for decision-making in the event of uncertainty [13]. Then, it follows the assessment of the impact (consequences) if the risk event becomes a reality. In this case, again, one or several risk factors (drivers) of potential consequences are identified. The impact probability *Pi* is determined in a way similar to the risk event definition. The model features another parameter, the total loss *Lt*, which is the loss that will occur if a risk event and the impacts are realised. The total loss may be expressed as a loss in time, in working days, in monetary terms (EUR), or in quality (e.g., the number of poor or substandard products).

The expected loss *Le* can be calculated according to Eq. (1) [10]:

$$Le = Pe \cdot Pi \cdot Lt \qquad (1)$$

In step 4, the criticality of the risk in question needs to be assessed separately from the qualitative and quantitative risk assessment and the total loss. We can use the calculation of the criticality level in the table of critical success factors, which is explained in detail in [9, 12]. In the proposed risk management model, we can use the risk map [10] shown in **Figure 3**.

A risk map is a diagram in which risk likelihood is on the y-axis and represents a product of the probability of risk event occurrence and the probability of risk impact (*Pe* • *Pi*), while the total loss *Lt* is on the x-axis. The threshold line of losses divides the surface of the diagram into two parts: the upper part above the threshold with the field of critical risks (Risk 1), which will have to be addressed by adopting adequate measures and the lower part below the threshold with the field of

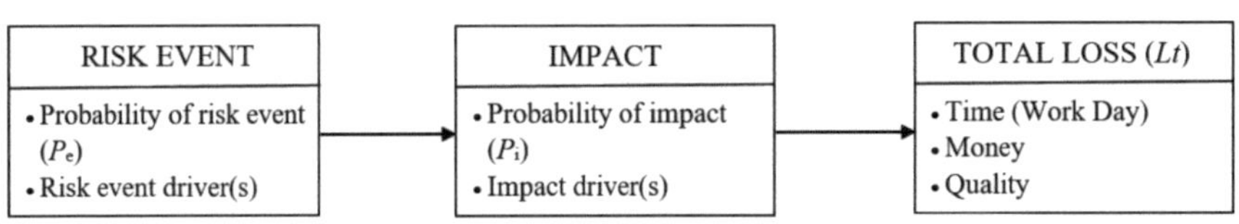

Figure 2.
Standard risk model.

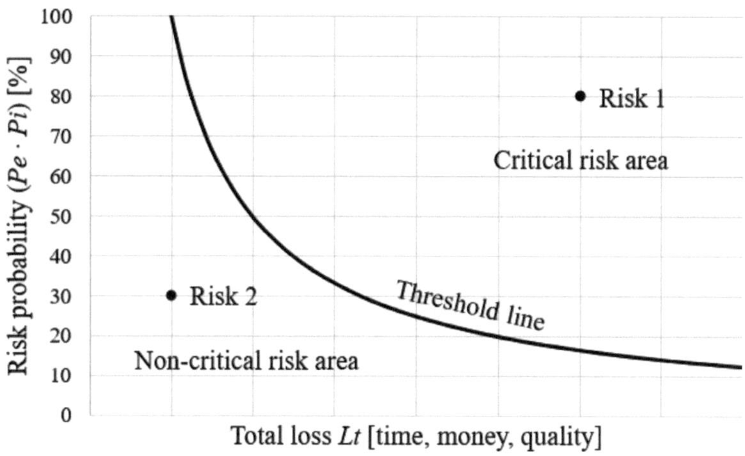

Figure 3.
Risk map.

noncritical risks (Risk 2), which are only identified and monitored, and measures are taken only if needed.

The threshold line of the expected losses is defined by Eq. (2) [10]:

$$Pe \cdot Pi = \frac{Le}{Lt} \tag{2}$$

Le in Eq. (2) represents the selected level of expected loss which is defined by the project team under consideration of the circumstances. It represents the value up to which the company is prepared to risk and accept the loss.

4. An example of infrastructure project risk management

The example of erecting a reservoir for a hydroelectric power plant (HPP) on the Lower Sava River [14] with a nominal power of 47.4 MW will be presented in the following. The HPP is of an impoundment facility type, with an arrangement of three vertical power units (double-regulated vertical power plant with a Kaplan turbine) with a nominal flow of 500 m^3/s with five flow-through fields and an average annual production of 161 GWh. The test operation of the HPP was foreseen for October 2017.

The HPP has a belonging reservoir with an anticipated 19.3 million m^3 of water on a surface area of 3.12 million m^2.

The planned goals for the erection of the reservoir for the HPP were as follows:

- A reservoir with the belonging infrastructure.
- Development of water infrastructure and state-regulated and local infrastructures on the influence area of energy utilisation of the river's water potential.
- A reservoir with high-water dams, drainage ditches, and other corresponding site development facilities.
- Treatment and maintenance of water infrastructure intended for preserving and regulating the quantities of water on the influence area of energy utilisation of the river's water potential.

- Running, maintaining, and monitoring the status of the water infrastructure intended for protection against detrimental effects of water on the influence area of energy utilisation of the river's water potential.

- Implementation of extraordinary measures during periods of increased hazard levels due to the detrimental impact of the waters on the influence area of energy utilisation of the river's water potential.

- Maintenance of water areas and acquired areas on the influence area of energy utilisation of the river's water potential.

- Providing sufficient quantities of water on the influence area of energy utilisation of the river's water potential.

- Flood safety for populated areas, protection of agricultural areas and forests, flood irrigation, and firewater catchment.

- Development of roads and other infrastructure.

- Passage for water organisms, spawning grounds, and other habitats and protection of landscape and cultural heritage.

- Recreational areas and cycling paths.

- Sediment depositions.

The main stakeholders involved in the implementation of the HPP reservoir are as follows:

- Investor with co-investors

- Contractor for project preparation and management

- Contractors for the execution of works

- The Government of the Republic of Slovenia, ministries with their bodies, and administrative units

- Other stakeholders (local communities, inhabitants, landowners, and pressure groups).

The investment value of the project amounted to EUR 140 million.

The contractor appointed a project team for the preparation and management of the project of the erection of the HPP reservoir. The project manager and team members received the following assignments: preparation of technical documentation, preparation of works, acquisition of lands, and maintenance and supervision of the entire project. Contractors for the execution of works were hired for the execution of individual activities.

A project of the erection of a reservoir for this hydroelectric power plant was selected because this was a big and important infrastructure project in Slovenia. This project is especially suitable for presentation of the proposed method of risk management due to its size and intervention in space, and the authors helped the contractor by project preparation especially in creating a project management plan.

4.1 Content and project timeline

The project team broke down the project's work content according to the WBS principle into the following phases: project preparation, designing, acquisition of permits, call for tender for the reservoir, dam house erection, and reservoir erection. For each phase, the team defined the necessary activities and linked them to a project network diagram. The network diagram links 242 activities. This is relatively little given the scope of the investment; however, the timeline here is only meant for the management of the investment and not for the operative management of the works of the project. The contractors prepared their own detailed timelines for the operative execution of the works of the project, which were fully harmonised with the project's timeline.

A project time analysis revealed that 1928 days are needed for the execution of the project of the erection of the HPP reservoir, with the beginning of the project scheduled for 1/3/2012 and the completion for 10/6/2017. The term of completion is very important, since the test operation of the HPP depends on it. **Figure 4** shows the project's timeline, wherein only the activities of the first phase are indicated.

4.2 Project's risk analysis

In the continuation, a method of use for risk management tools is shown using an example of an infrastructure project. In compliance with the method of **Figure 1**, an Ishikawa diagram of project risks was first drawn up, in which the key risk factors (groups) in this project have been identified: environment, contractor, client, Government of the Republic of Slovenia, and project execution. Possible risks in the project have been identified for individual risk groups (**Figure 5**).

The use of the Ishikawa diagram proved a very efficient tool in our case, since the team members had already used it in the quality management. The team members highlighted those risks that are most likely to occur in this project and inserted them in the prepared table template of critical success factors from the MS Project software according to [9, 12]. The probability of a risk event occurrence and a probability of consequences were assessed for each activity according to the Likert five-point scale (1–5), and a risk rate for the activity was calculated. It is marked with as indicated (colour indicators: red, high; yellow, medium; and green, low risk rate). **Figure 6** shows part of the project's risk analysis for the activities of the first phase (WBS group), which is project preparation.

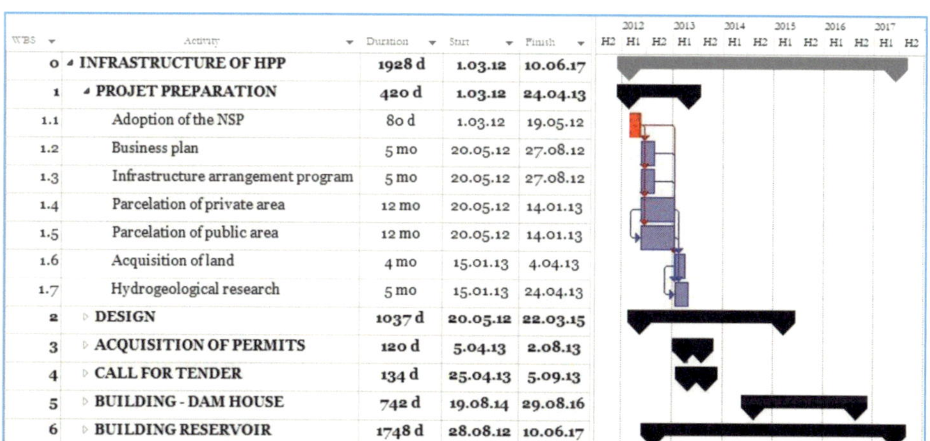

WBS	Activity	Duration	Start	Finish
0	INFRASTRUCTURE OF HPP	1928 d	1.03.12	10.06.17
1	PROJET PREPARATION	420 d	1.03.12	24.04.13
1.1	Adoption of the NSP	80 d	1.03.12	19.05.12
1.2	Business plan	5 mo	20.05.12	27.08.12
1.3	Infrastructure arrangement program	5 mo	20.05.12	27.08.12
1.4	Parcelation of private area	12 mo	20.05.12	14.01.13
1.5	Parcelation of public area	12 mo	20.05.12	14.01.13
1.6	Acquisition of land	4 mo	15.01.13	4.04.13
1.7	Hydrogeological research	5 mo	15.01.13	24.04.13
2	DESIGN	1037 d	20.05.12	22.03.15
3	ACQUISITION OF PERMITS	120 d	5.04.13	2.08.13
4	CALL FOR TENDER	134 d	25.04.13	5.09.13
5	BUILDING - DAM HOUSE	742 d	19.08.14	29.08.16
6	BUILDING RESERVOIR	1748 d	28.08.12	10.06.17

Figure 4.
Timeline for the erection of the HPP reservoir.

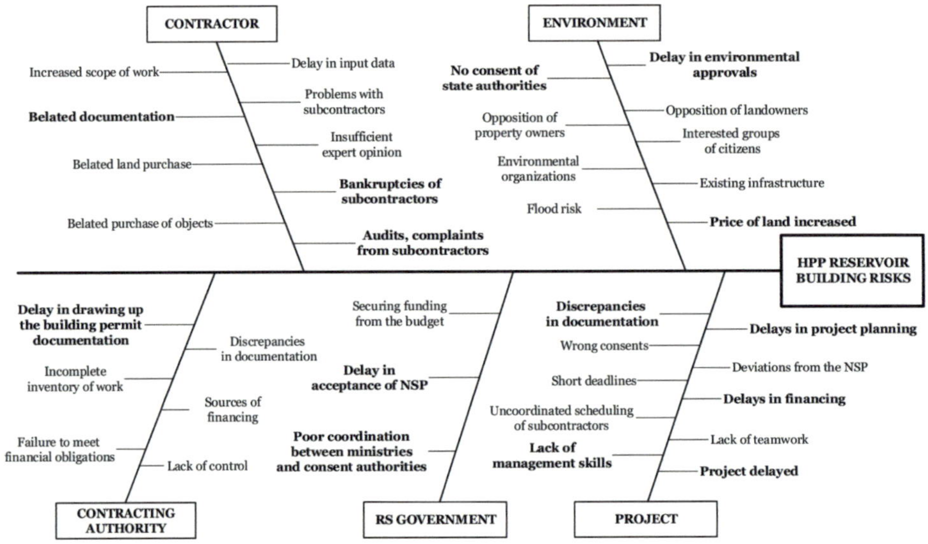

Figure 5.
Ishikawa diagram of the HPP project risks.

WBS	Activity	Causes	Consequences	Event probability	Consequence probability	Rate of risk
0	◢ INFRASTRUCTURE OF HPP					
1	◢ PROJET PREPARATION					
1.1	Adoption of the NSP	Delay in acceptance of NSP	No consent of state authorities	5	5	●
1.2	Business plan	Discrepancies in documentation	Delays in project planning	3	3	●
1.3	Infrastructure arrangement program	Belated documentation	Delays in project planning	3	3	●
1.4	Parcelation of private area	Opposition of landowners	Delays in project planning	3	3	●
1.5	Parcelation of public area	Opposition of property owners	Delays in project planning	1	1	●
1.6	Acquisition of land	Opposition of landowner and propery owners	Price of land and real estate increased	3	3	●
1.7	Hydrogeological research	Delay in environmental approvals	Project delayed	2	2	●
2	▷ DESIGN					
3	▷ ACQUISITION OF PERMITS					
4	▷ CALL FOR TENDER					
5	▷ BUILDING - DAM HOUSE					
6	▷ BUILDING RESERVOIR					

Figure 6.
Risk analysis in the MS project software (part).

Based on the risk analysis of all project activities, the project team established that eight activities have a very high risk rate, which is why they decided to analyse these risks in more detail using the standard risk management model. For each of the eight high-risk activities, the project team determined the probability of a risk event *Pe* occurrence and a probability of consequences *Pi* and calculated the overall risk probability. The total loss *Lt* and the expected loss *Le* calculated using Eq. (2) were assessed.

The majority of risks in question result in a delay in the project and consequently in the launch of the HPP test operation. Our assessment of losses was based on data that indicated 1 day of interrupted operation of such an HPP means a loss of income of 17,600 EUR/day. Calculations for the expected losses are given in **Table 2**.

As evident from **Table 2**, some risk-related losses refer to monetary losses and others to time losses, which is why the risks related to monetary losses in terms of

No.	Activity	Risk description	Designation	P_e	P_i	$P_e \times P_i$	L_t [10^3 € or month]	L_e [10^3 € or month]
1	Adoption of national spatial plan—DPN	Delay in adoption	T1	0.9	1	0.9	€320	€288
2	Adoption of national spatial plan—DPN	Delay in adoption	T2	0.9	1	0.9	6 months	5.4 months
3	Reservoir plan	Extra costs	T3	0.7	0.5	0.35	€50	€17.5
4.	Reservoir plan	Delay in execution	T4	0.8	0.8	0.64	6 months	3.8 months
5	Tender	Conditions not met	T5	0.9	0.7	0.63	2 months	1.3 months
6	Acquisition of buildings and land	Opposition of owners	T6	0.7	0.8	0.56	€200	€112
7	Acquisition of buildings and land	Opposition of owners	T7	0.7	0.6	0.42	5 months	2.1 months
8	Existing roads	Damage to existing roads	T8	0.8	0.9	0.72	€750	€540

Table 2.
Evaluation of losses in huge project risks relating to the erection of the HPP reservoir.

extra costs (**Figure 7**) are shown in the risk map separately from the time losses due to delays (**Figure 8**).

In the risk map, in which the risks related to extra costs (**Figure 7**) are shown, four risks are identified: of those, three (T1, T6, and T8) are critical, while the T3 risk belongs to the group of noncritical ones. The project team subsequently prepared adequate measures for the critical risks to prevent or mitigate the consequences if a risk event were realised.

In the risk map, in which the risks related to delays (**Figure 8**) are shown, four risks are identified as well: of those, two (T2 and T4) are critical and two (T5 and T7) belong to the noncritical risks. Again, the project team prepared adequate measures for the critical risks to prevent or mitigate the consequences if a risk event were realised.

To illustrate preparation, analysis, and assessment of further measures to mitigate the risk consequences, the T1 risk was selected, i.e., the adoption of the DPN (state spatial plan). This is the first activity in the project having a high and critical risk (particularly due to the fact that it can delay the project's execution and due to the extra costs incurred). The state spatial plan is adopted by the Government of the Republic of Slovenia upon a proposal from the ministries and in conformity with other bodies issuing permits for the placement of infrastructure projects into the space. Local communities actively participate in this process. To mitigate the consequences of the T1 risk (delay in the adoption of the DPN), the project team prepared a plan of measures in three iterations, as shown in **Table 3**.

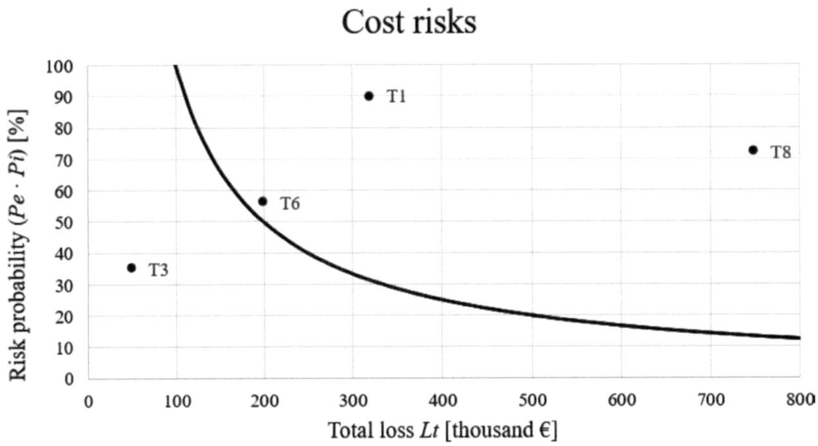

Figure 7.
Map of project risks expressed by costs.

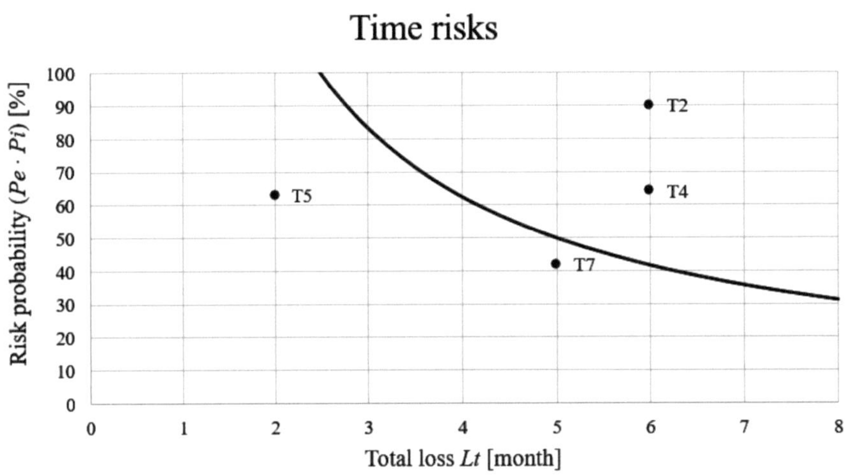

Figure 8.
Map of project risks expressed as delay in time.

For the anticipated measures, the project team assessed the probabilities of a risk event occurrence, a probability of impact after the adoption of a measure and the total and expected loss. Ten percent of the anticipated loss of income was determined by the project team as the value of the total loss. The calculations and results for all three iterations of measures are shown in **Table 4**.

Based on the results from **Table 4**, a risk map for the T1 risk was drawn up (delay in adopting a DPN), which is shown in **Figure 9**.

In the risk map in **Figure 9**, the threshold line denotes a still acceptable value of loss of EUR 100,000, which the project team considers to be the maximum tolerable value. T1 represents the starting situation, and there are no extra measures except warnings to the Government of the RS to start preparing a DPN for the erection of the HPP reservoir. The expected delay here will be 6 months, and the expected loss incurred by the client due to a delay in the scheduled start-up of the HPP is EUR 288,000. Point T1.1 represents a point of risk T1 in the risk map after a measure is adopted, with which the ministry responsible for infrastructure would appoint a co-ordinator to co-ordinate the preparation of the DPN. Still, a delay of 5 months is expected, while the expected loss in this case would be reduced to EUR 180,000, which still means that the risk is a critical one. Additional suggested measures, with

Risk	Risk event driver	Prevention plan	Impact driver	Contingency plan
T1	Government of the RS is often late in adopting a resolution on preparing a DPN	Client should warn the government of the consequences of delay in adopting the resolution on the preparation of a DPN	Minimum 6-month delay in preparation of documentation	A loss of income totalling EUR 3.2 million is anticipated
T11	Discrepancy in the work of ministries in the preparation of a DPN	Suggest that the Ministry of Infrastructure assumes co-ordination	Minimum 5-month delay in preparation of documentation due to discrepancies within ministries	A loss of income of €3 million is anticipated, despite the appointment of a co-ordinator for the preparation of a DPN
T12	Comments of municipalities on the proposed DPN	Mismatched comments between municipalities	Mismatched comments between municipalities represent a further 4 months of delay	Comments of municipalities get matched, yet a further €2.5 million loss of income is still anticipated

Table 3.
Plan of measures to reduce consequences of risk in the preparation of the DPN.

Risk	Probability of risk event *Pe*	Probability of impact *Pi*	Total probability *Pe* × *Pi*	Total loss *Lt* (thousand EUR)	Expected loss *Le* (thousand EUR)
T1	0.9	1	0.9	320	288
T1.1	0.6	1	0.6	300	180
T1.2	0.4	0.8	0.32	250	80

Table 4.
Results of calculated impacts of the suggested measures to mitigate risk T1.

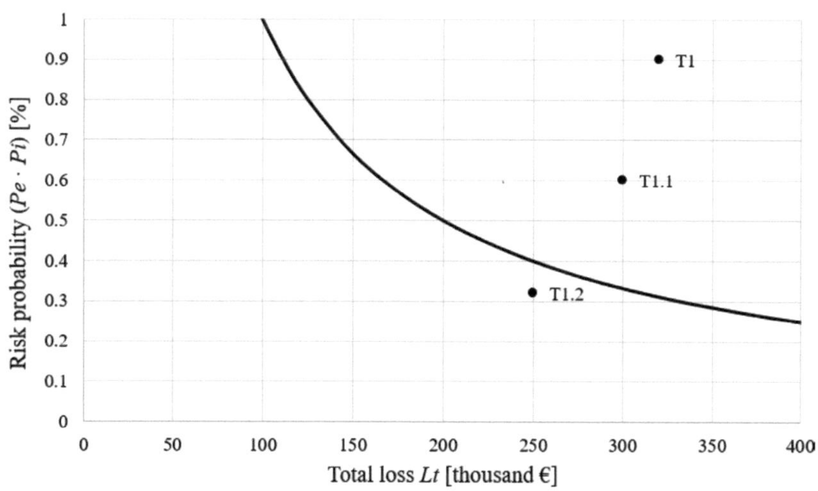

Figure 9.
Risk map after the introduction of anticipated measures to mitigate risk T1.

which the municipalities on the territory in which the HPP reservoir will be erected, would co-ordinate among themselves and could render risk T1 noncritical, because point T1.2 lies below the threshold line of expected losses. A delay of 4 months is still expected, yet the expected loss is EUR 80,000, which is less than the maximum threshold value of EUR 100,000 the project team had determined for this risk.

In the project execution phase, the activities of the project must be closely monitored, and attention should be paid to the time of risk event occurrence. The project team can determine risk indicators [12] that remind them of points in time when a potential risk event might or could be expected to occur. It is not enough only to introduce measures, the situation should be constantly monitored and additional measures adopted to mitigate the impacts of risk events that occur. In the case of the erection of the HPP reservoir, the project team also constantly monitored the risk management activities and adopted adequate measures as required. The management of risks in the project in question was ultimately successful, since the HPP started operating on schedule and according to the timeline.

It is important to note that once the project was completed, the project team made a thorough analysis of their risk management strategy and identified those solutions that proved effective and successful and that would be worth using in similar projects in the future, as well as ineffective solutions that should be avoided in future projects.

5. Conclusions

Risk management is an important field of knowledge that is an integral part of [any] efficient project management. It is important that risk management be completely integrated into other areas of project management. The paper dealt with the risk management in infrastructure projects, which, compared to other projects (e.g., product development or IT projects), involve considerably more impact factors related to the environment and that are included in the process of planning and management of such projects.

The paper outlined the methods and tools that project management can use in project planning and management. The following methods are of particular importance for managing of an infrastructure project: an Ishikawa diagram for identification of potential risks; a table of critical success factors that identified risks to individual activities and classifies the risks of the activities as high-, medium-, and low-risk; a standard risk model that serves to determine expected losses in time, money, and quality; and finally, a risk map that classifies a risk as a critical or noncritical risk. The risk map can be used to analyse how the anticipated measures could work to reduce the critical nature of the risk.

The above-indicated methods have been successfully tested in the erection of the HPP reservoir. The project represented an important instance of interference in the space, even more than the placement of the HPP itself. It has been proved that the key risks in this project were those risks on which neither the investor nor the contractors have any influence. In our case, this was the integration of the building into the space and problems relating to the preparation of the DPN, which is crucial for further planning and subsequent project management. Risks also appeared in the acquisition of the land and in respect of the requirements demanded by parties granting the relevant permits, by the state, the groups with special interests, and pressure groups (conservationists).

The use of the proposed extended model for the identification, assessment, prioritisation, and management of risks proved highly successful in the HPP project. The table of critical success factors also proved very successful. It was created using the MS Project software that was also used for the planning and monitoring of the project. This integration allowed the project team to have the risk management data available in the same tool as other project management data, which proved to be particularly efficient in monitoring the execution of the project.

What was new here was the use of the standard risk model, with which the project team could (as with the critical success factor table) identify and quantify the importance of each risk and assess the expected loss. The advantage of using the model is its simplicity; its key drawback, however, lies in the fact that the result depends on the accuracy of team members' assessments of the probability and total loss factors. Nevertheless, this drawback did not prove a substantial disadvantage. The risk map, with which risks are classified as critical and noncritical, proved to be a very important tool. Determining the threshold line of acceptable losses could appear as a problem, as it is based on a subjective assessment of the team members. Also, the possibility of checking the impacts of the foreseen measures adopted for the most critical risks is important; yet as it turns out, there is often a lack of motivation among team members, and they prefer, instead, to simply follow their intuition.

The execution of the project in question revealed that infrastructure projects are considerably more demanding than other projects in terms of risk management. As a rule, stakeholders from the wider environment have to participate in such projects.

In any follow-up (work, analysis, research), it would be important to consider how to support the subjective determination of the data for the use of the standard model by means of decision-making methods in cases where there is an element of uncertainty present.

The results of the proposed extended model for managing the risks of this infrastructure project will be a great help to project managers who will carry out similar projects in the future.

Acknowledgements

The authors gratefully acknowledge the financial support of the Slovenian Research Agency (research core funding no. P2-0270, agreement no. 1000-15-0510).

Author details

Lidija Rihar, Tena Žužek, Tomaž Berlec and Janez Kušar*
Faculty of Mechanical Engineering, University of Ljubljana, Ljubljana, Slovenia

*Address all correspondence to: janez.kusar@fs.uni-lj.si

References

[1] Iqbal S, Choudhry RM, Holschemacher K, Ali A, Tamošaitiene J. Risk management in construction projects. Technological and Economic Development of Economy. 2015;**21**(1):65-78. DOI: 10.3846/20294913.2014.994582

[2] Choudhry R, Iqbal K. Identification of risk management system in construction industry in Pakistan. Journal of Management in Engineering. 2013;**29**(1):42-49. DOI: 10.1061/(ASCE)ME.1943-5479.0000122

[3] Hameed A, Woo S. Risk importance and allocation in Pakistan construction industry: A contractor's perspective. KSCE Journal of Civil Engineering. 2007;**11**(2):73-80. DOI: 10.1007/BF02823850

[4] Yafai KN, Hassan JS, Balubaid S, Zin RM, Hainin MR. Development of a risk assessment model for Oman construction industry. Jurnal Teknologi. 2014;**70**(7):55-64. DOI: 10.11113/jt.v70.3579

[5] PMBOK Guide. A Guide to the Project Management Body of Knowledge. 6th ed. Newtown Square, Pennsylvania, USA: Project Management Institute, Inc.; 2017

[6] Managing successful projects with PRINCE2, 6th ed, London: AXELOS; 2017

[7] DOD Risk Management Guide for DOD Acquisition. 6th ed. USA: Department of Defence; 2006

[8] Dobie Colin A. Handbook of Project Management. Australia: Allen Unwin; 2007

[9] Rihar L, Berlec T, Kušar J. In: De Felice F, Petrillo A, editors. Theory and Application of Cognitive Factors and Risk Analysis, New Trends and Procedures. Rijeka, Croatia: InTech; 2017. pp. P63-P86. DOI: 10.5772/intechopen.68398

[10] Smith GP, Merritt MG. Proactive Risk Management. New York, USA: Productivity Press; 2002

[11] Hilllson D. Using a risk breakdown structure in project management. Journal of Facilities Management. 2003;**22**(4):85-97. DOI: 10.1108/14725960410808131

[12] Kušar J, Rihar L, Žargi U, Starbek M. Extended risk-analysis model for activities of the project. Springer plus. 2013;**2**:227

[13] Götze U, Northcott D, Schuster P. Investment Appraisal, Methods and Models. Berlin Heidelberg: Springer-Verlag; 2015

[14] Hydroelectric Power Plants on the Lower Sava River. Available from: http://www.he-ss.si/eng/ [Accessed 07-12-2018]

Chapter 2

A New Mode of HSE Risk Management for Construction Projects

Yueting Hu

Abstract

The HSE case, developed by Shell Co., is a quite famous HSE risk management tool for construction project, but too much content makes it difficult to compile it before the start of a project. Even though the HSE case is finished, it is hard to have so much content to be mastered by the persons concerned in so limited time just before the start of a project; it is thus called "the case sits on the shelf." In order to make such a tool much more practical, we adapt it according to its application environment, and the mode of "Two Documents and One Checklist" is thus formed. The mode contains two documents and one checklist. One of the documents is a relatively static document, named "work-post HSE guide," which is designed to manage the relatively static risks; the other document is a changing document, named "project HSE plan," which is designed to manage the changing risks. Both documents are designed mainly to guide workers to work or operate in a standard and safe manner. The checklist is designed to verify whether the condition of the workplace such as machines, equipment, tools, and so on is safe or not. Owing to the feature of the mode of "Two Documents & One Checklist," it is not only quite easy to compile but also very convenient to apply in daily work by eliminating the problems that appear in the HSE Case.

Keywords: HSE risk, construction projects management, front-line organization

1. The features of risk management of construction projects

The risk management can be applied in many fields wherever risks exist. In the field of health, safety, and environment management, it is called HSE risk management; in the field of finance, there is finance risk management and so on.

Just as for safety risk management, different objects need different risk management modes because of their different features; therefore, many different risk management modes are developed to manage different kinds of things. As for construction project, its obvious feature is the change; the change may cover personnel, machinery and equipment, raw materials and products and semi-finished products, technology, environment (natural environment and social environment), etc. Therefore, the mode for the safety risk management of construction project should meet its changing feature.

Safety case and HSE case are two typical modes applied for the safety and HSE risk management of construction projects. Although they are used widely around the world, many problems aroused during our application of the method in our daily HSE management of the construction projects.

2. The problems and the objectives

Although we pay more attention to accident prevention from the state to our corporation, we own little effective safety risk prevention method before. What we have done is to learn the lessons after the accident happened instead of taking prevention beforehand. We have been exploring a suitable method to apply the risk management theory to the daily production and operation of grassroots organizations since the introduction of HSEMS [1]. When we got to know the HSE case [2], we found it was a good method to prevent accident beforehand at least in theory, but many problems aroused during our application of the method in our daily HSE management of the construction projects in the frontline organizations.

HSE case is a comprehensive document for risk prevention [3, 4]. It is rich in content and covers a wide range, which however will not only increase the preparation workload but also affect its implementation. To prepare a HSE case, it is necessary to comprehensively identify HSE hazards and assess their risk, develop corresponding risk prevention measures, and establish documents in writing before a project started [5]. Since all these work must be done before the project is started, problems such as tight schedule and burdensome task may be encountered, leading to failure in compiling such kind of a huge document, let alone its quality. On the other hand, as HSE case is rich in contents, there is usually no enough time to organize a process to educate relevant personnel in the project preparation phase. Even if the education process is implemented, the effect will always be too poor due to its too many contents,. In view of the above problems, some companies only regard HSE case as "a letter of guarantee" and submit it to relevant stakeholders, emphasizing their concern for project HSE risk prevention while downplaying relevant education. As a result, even the companies within Shell Co. once internally dubbed HSE case as "the case sits on the shelf." In a word, although there are many problems mentioned above, the key problem is its too much content of the HSE case. Because of too much content, it is difficult to compile such a huge document within the limited time just before the beginning of a project especially for the frontline organizational persons, let alone train the workers with it in such a short time.

According to the above analysis, the main objective is to reduce its content and to make the document of HSE case a bit simple in order to compile and apply it in daily HSE management of the construction projects in the frontline organizations smoothly. Based on the analysis, a new safety risk management mode for construction project, called the HSE-TDOC (two documents and one checklist), was developed in 2001 [6] and was modified in 2007 according to the problems met in practice [7]. The underlying principle and application of the model, as well as the document structure and steps of compilation, were described in the following, through which the model is well explained.

3. The methodology used to develop the HSE-TDOC

HSE case is a kind of HSE risk management document developed to enhance the project HSE risk prevention capabilities. Its biggest advantage lies in the organic integration of HSE risk management theories and the practice. It applies risk management theories to effectively guide actual HSE risk management, especially the project HSE risk management.

Based on the problems met in the application of HSE case, a new safety risk management mode, named the HSE-TDOC (two documents and one checklist), was developed for construction project, and it will be introduced in this section.

3.1 The analysis of the HSE case

In fact, HSE case has become not only "the case sits on the shelf" but has been questioned by some experts and scholars for its way of managing operational risks. For example, in the co-authored paper "Integrating Safety Management Through the Bowtie Concept—A move away from the Safety Case Focus," Australian scholars Acfield et al. [8] believed that HSE case is applicable to managing risks arising from changes in the project or activities, and it should not be used to manage operational risks.

Through the study of risk management theories and systematic analysis of various risks, we believe that the risks encountered in practice can be roughly divided into two types, i.e., "relatively stable" risks and "changing" risks [9]. "Relatively stable" risks have two characteristics. The first feature is just as the name implied that they are relatively stable, e.g., in oil or gas well-drilling industry, the blowout risks while drilling is a "relatively stable" risk. As long as there are no great changes to the work object, process and technology, equipment, and facilities, such risks will keep being stable; whenever, wherever, or whoever has an oil or gas well drilled, the risk of well blowout will always exist. Because the underground high-pressure fluid layer may be meted while drilling, the well blowout will happen if the preventive measures are null and void. The second feature is that they are specialty-related. The blowout risks, for example, may only exist in business areas related to the underground high-pressure fluid layer, such as drilling, logging, and workover, while in other unrelated fields such as refining and chemical and transport, blowout accidents are impossible.

Such "relatively stable" risks are actually the so-called operational risks by Acfield et al. [8]. They are often called as conventional risks or conventional operational risks, because they generally occur in the course of conventional operations. Conventional operations refer to those operations with relatively fixed work contents and environment that can be carried out according to preset procedures, which is also named standard operating procedure (SOP). Therefore, risks arising from conventional operations can be prevented by complying with the corresponding operating procedures, working procedures, SOP, etc. that aim to regulate the behavior of operators. As conventional risks are relatively stable, and the measures to control them are also stable, there is no need to manage such kind of risks based on project-specific HSE case which is changed from one project to another.

Compared with conventional risks, unconventional risks have unique characteristics. Firstly, unconventional risks are changing. Prevention of such risks requires pertinent measures considering many associated factors, not like conventional risks which can be prevented by developing operating procedures, working procedures, SOP, etc. Such risks are called as "change risks" by Acfield et al. [8]. Secondly, although there are many types of unconventional risks (risks arising from unconventional operational activities and changes), the total amount of is quite less than that of conventional ones. Thirdly, unconventional risks are unrelated to specialties and may exist in any field. Because of the above characteristics, unconventional risks are more suitable for HSE case. As different projects may encounter different unconventional risks, it is necessary to identify, evaluate, and develop control measures of unconventional risks from one project to another.

3.2 The development of HSE-TDOC

As the abovementioned, since one kind of risk is specialty-related, a relatively stable document can be developed to meet such kind of needs. Therefore, according to different kinds of specialties, we develop relatively stable HSE guidance which are specific to the specialties or work post. Preparing HSE guidance may take a lot of time and effort because contents of such risks management are quite wide. However, seeing that such kind of risks is relatively stable and there is no deadline for completion, HSE guidance can be used for a long time once completed.

Another type of risk is changing risks which we call unconventional risks. Unconventional risks refer to the risks other than conventional risks (**Figure 1**) [10]. They include not only the risks arising from a variety of unconventional operational activities (operational activities that cannot be carried out according to established procedures due to changes in job contents, environment, etc.) such as risks arising from hot work, excavation, work at height, etc. but also the risks brought about by changes, such as the risks brought about by changes in personnel, equipment, raw materials (finished and semi-finished products), process and technology, environmental factors (natural environment and social environment), etc.

As mentioned earlier, a large amount of contents on the prevention of conventional risks has been formed into a new document called specialty-specific HSE guidance, resulting in much of the content of the former HSE case being stripped out, with the contents on the management of unconventional risks (including project emergency management) being included in the HSE case. In this case, we take a new name for HSE case, i.e., project-specific HSE plan (HSE plan). Actually, HSE plan is a downsized HSE case. As different projects may encounter different unconventional risks, each project should be prepared with its own HSE plan according to its characteristics. Generally, a project may not encounter too many unconventional risks, so the HSE plan is usually easy to prepare and communicate. Furthermore, as the unconventional risks are the risks that exist in the project but are not included in HSE guidance, they are the ones known as the new additional risks of the project in practice; therefore, HSE plan is also called the document for managing additional risks of the project.

As the abovementioned, HSE plan is a downsized HSE case, so the HSE plan will be prepared and applied just as the HSE case. To prepare a HSE plan, before a project started, people concerned should go to the worksite of the project to conduct site surveys and collect relevant information and data in order to comprehensively identify and assess HSE hazards, develop corresponding risk prevention measures, and establish documents in writing. Certainly, those that have been already managed by the HSE guidance will not appear in this document. The HSE plan mainly deals with the risks caused by the change of personnel, machinery and equipment,

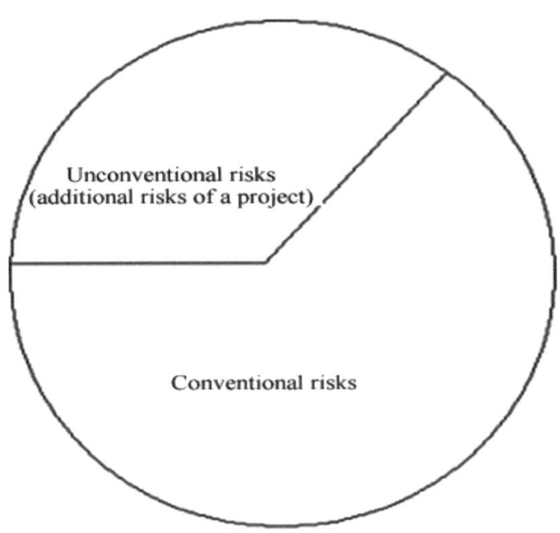

Figure 1.
Conventional risks and unconventional risks.

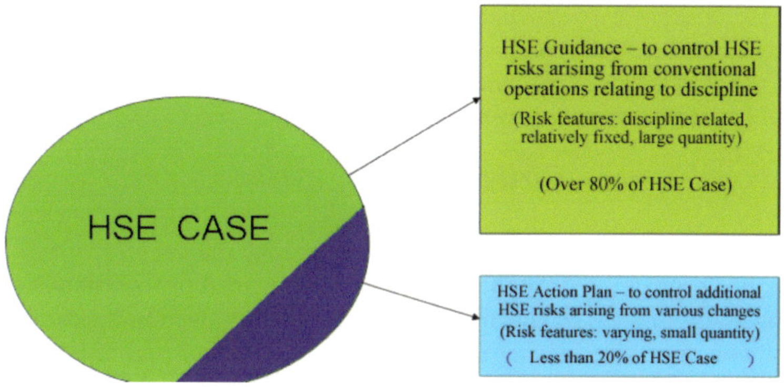

Figure 2.
Relations between HSE case and "two documents."

raw materials and products and semi-finished products, technology, environment (natural environment and social environment), etc. For example, a grassroots team which usually works in plain areas may occasionally go to mountain areas to work there. Due to changes in the natural environment, the grassroots team may encounter mountain torrents, landslides, and other natural disasters that will not happen in plain areas before. No matter what kind of unconventional risks, if they are needed to be controlled, they should be included in the HSE plan as long as they appear in the project.

As mentioned in the "two documents and one checklist" previously, the "two documents" are essentially the results of dividing HSE case based on the nature and characteristics of HSE risks (**Figure 2**).

4. The "two documents and one checklist" risk management mode

In this section, the mode of HSE-TDOC will be introduced thoroughly. First is the overview of HSE-TDOC, and then its content and compilation and application will be followed; at the end, its function and effect will be mentioned too.

4.1 Overview of HSE-TDOC

HSE-TDOC refers to the specialty-specific HSE guidance (HSE guidance), project-specific HSE plan (HSE plan), and position-specific HSE checklist (HSE checklist) [11]. Among them, the HSE guidance may have much content, while the HSE plan may either have much content or just a few pages depending on the project.

4.1.1 Specialty-specific HSE guidance (HSE guidance)

HSE guidance is a guiding document which is used to reduce the HSE risks arising from discipline-related conventional operations to the ALARP level through risk management. Through the risk management process, countermeasures against the HSE risks to be managed are developed. Then these countermeasures that are distributed to relevant positions with written records are kept. After being reviewed by the competent department (personnel), the written records are compiled into the HSE guiding document specific to the discipline.

As the HSE risks arising from discipline-related conventional operations are relatively stable, there would be no change in the corresponding prevention and control measures as long as no change occurred to process, technology, equipment, facilities, etc. If there are risks that resulted from temporary changes, control of such risks should be carried out by means of the HSE plan. Certainly, if there are risks that resulted from forever changes, the HSE guidance should be modified. Therefore, the HSE guidance is relatively stable.

4.1.2 Project-specific HSE plan (HSE plan)

The HSE plan is a document intended to control the risk result from all kinds of change. It is prepared before the project/activity started in accordance with the risk management process by the operating personnel involved in the project/activity after site surveys, considering the "person, machine, material, method, and environment, etc." influencing factors and changes thereof. It shall be reviewed and approved by the competent department (personnel) upon completion.

The HSE plan is prepared on the basis of the HSE guidance. It is a supplement to HSE guidance and covers the prevention and control measures against HSE risks of the project/activity that are not included in the HSE guidance. Together, the HSE plan and the HSE guidance are documents to control all the identified risks of the project/activity. Comparison of the characteristics of the two is shown in **Table 1**.

4.1.3 Position-specific HSE checklist (HSE checklist)

The "one checklist" in the "two documents and one checklist" refers to the position-specific HSE checklist. It is a form designed according to a scientifically reasonable route (order) to prompt inspection personnel to pay more attention to critical parts or vulnerable components of field hardware facilities that are used or controlled by employees on each position, such as tools and machines, equipment, etc., in order to improve the efficiency of discovering hidden dangers. Each position has its own corresponding checklist. Through using the checklists of all positions, all the field hardware equipment, facilities, tools, and machines can be fully inspected to ensure that they are in a safe condition. The position-specific HSE checklist can not only ensure full inspection and effective management on field objects that are in an unsafe condition but also improve the efficiency of safety inspection.

Features	**HSE guidance**	**HSE plan**
Object	Post or discipline, compiled by post as much as possible	Project, activity
Compilation time	No strict restrictions on compilation time. At best, it should be compiled when such organization is established	Before the project/activity commencement (strictly restricted)
Features	Rich contents, relatively fixed, available for long-term use	Simple content, a temporary document of "one case one meeting"; the plan is annulled after the project is completed
Application	Reference in daily work, the main training data for centralized study and training. Via the daily or centralized study and training, the employees' professional qualities will be enhanced, and the conventional risks will be effectively prevented	Before the project/activity commencement, education is carried out for all employees involved in the project/activity, so that they will know additional conventional risks and know how to implement corresponding prevention measures. In this way, additional conventional risks of the project/activity can be prevented

Table 1.
HSE guidance vs. HSE plan.

4.2 The content and compilation and application of HSE-TDOC

4.2.1 The HSE guidance

4.2.1.1 The content of HSE guidance

The content of HSE guidance includes, but not limited to, the following contents; in other words, the main contents of the HSE guidance are as follows:

- Job qualifications
- Job responsibilities
- Standard operating procedures
- Patrol inspection and main contents
- Emergency response procedures
    - "Job qualifications" and "job responsibilities" are the basic requirements for a position developed by the personnel department based on the characteristics of the position. It is the responsibility of an employee to do a good job in his post. While meeting the qualifications for the position, an employee must also be aware of the responsibilities of the position. Therefore, job qualifications and job responsibilities are the most basic requirements that an employee must meet.
    - "Standard operating procedures" and "emergency response procedures" are the core content of the HSE guidance. With the deepening of standardized management in enterprises, basically each conventional operation is provided with standard operating procedures. But due to poor education on these procedures, they were not well understood by employees. So usually operations were carried out beyond the standard procedures, which are the main causes of most accidents. One of the functions of the HSE guidance is to have the employees understand the standard operating procedures by providing them in print to the employees. If necessary, employees may refer to the HSE guidance in advance to prevent nonstandard operations. Emergency response procedures are actually the operating procedures in a state of emergency. Since it keeps relatively stable once established and modified, we can analyze various emergencies that may occur to a specific position and incorporate the corresponding "emergency response procedures" into the HSE guidance specific to the position, so as to improve emergency response skills of employees.
    - "Patrol inspection and main contents" aims to ensure the overall safety of objects. First of all, set the main contents of inspection according to the characteristics of the tools, machines, equipment, facilities, and other hardware items used or managed by the employees on the post, especially critical parts and vulnerable components, based on the principle of territorial management. Then, devise an inspection route to find problems and hazards of the mentioned hardware items, aiming to ensure the overall safety of objects effectively. By the way, "patrol inspection and main contents" in the HSE guidance is just to govern the use of its HSE checklist.

Except for the "patrol inspection and main contents," all the rest are readily available. Incorporate them into the HSE guidance after modifications are made to relevant contents through risk management activities.

The revised position-specific HSE guidance (version II) mainly contains the basic requirements that employees should meet. Unlike version I, the contents on the management of conventional risks relating to the discipline in version II were incorporated into position-specific operating procedures and other relevant contents, facilitating the education on the HSE guidance.

In addition, we also recommend the use of "bowtie" model and other effective methods to reinforce the prevention of significant risks. Significant risks generally fall into the category of conventional risks. In addition to the normal prevention means such as operating procedures, we recommend using the "bowtie" model (**Figure 3**) to strengthen the prevention of significant risks. Depending on the circumstances, the "key tasks" and "key facilities" generated by the "bowtie" model should be allocated to relevant positions and included in the position-specific HSE guidance.

4.2.1.2 Compilation and application of the HSE guidance

The HSE guidance should be compiled considering the nature of grassroots organizations. For grassroots organizations of the same type, a consistent HSE guidance could be compiled since their disciplines and position settings are the same. The discipline-specific HSE guidance should be compiled usually by the enterprise or its subsidiaries. When compiling the discipline-specific HSE guidance, attention should be focused on the discipline and related activities and potential abnormalities and emergencies in the entire process of the project, followed by hazard identification, risk assessment, and development of appropriate risk control measures. The HSE guidance shall be reviewed and approved by the competent department.

Since the HSE risks arising from discipline-related conventional operations are relatively stable, risk prevention measures may stay the same as long as no change occurred to the process, technology, equipment, facilities, etc. Hence the HSE guidance is a kind of relatively stable document. It can be used as a working guide for relevant employees in their day-to-day work and can also be used as a resource for self-study. More importantly, grassroots organizations should regard the education on the HSE guidance as a management action that should be persisted for a long time, so as to enhance the staff's professional quality and risk prevention capability.

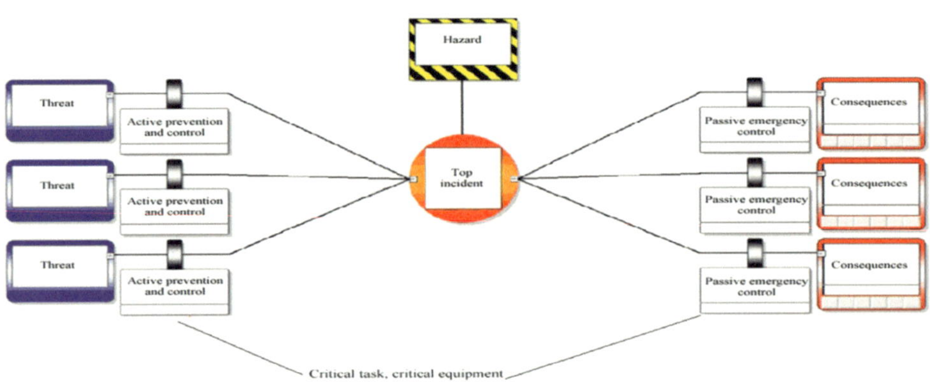

Figure 3.
Bowtie model for preventing significant risks.

As the HSE guidance has relatively fixed content, its compilation and education could be a part of day-to-day work, avoiding the problems (heavy compilation and education workload before the start of a project) and faced by HSE case.

All in all, through the education on the HSE guidance during daily work, relevant employees will understand the conventional risks and characteristics thereof related to their positions and know how to implement the specific measures to prevent these risks, thus effectively improving their ability to control the discipline-related conventional risks.

4.2.2 The HSE plan

4.2.2.1 Content of HSE plan

The content of HSE plan includes, but not limited to, the following contents; in other words, the main contents of the HSE plan are as follows:

- Project overview, worksite, and the surroundings
- Personnel and equipment
- Identification of additional hazardous factors and main risk warnings
- Risk prevention, mitigation, and
- Emergency response plan

    - "Project overview, worksite, and the surroundings" and "personnel and equipment" are set up to identify hazardous factors. For projects of mobile operations, changes may most likely happen to the project itself and the surrounding environment, personnel, equipment, and facilities, so the HSE plan focuses on these two parts for comprehensive identification of hazardous factors. Specifically, in the "personnel and equipment" part, potential risks are identified through analyzing changes (placement, shifts, etc.) in project personnel (especially those in key positions), and appropriate measures are developed. Regarding equipment and facilities, risks arising from frequent relocation and installation are considered, such as safety accessories lost, damaged, etc.

    - "Identification of additional hazardous factors and main risk warnings" and "risk prevention, mitigation, and control" are the focus of the HSE plan. When compiling the plan, the first step is to identify hazardous factors through analyzing the "project overview, worksite, and the surroundings" and "personnel and equipment." On this basis, the next step is to refer to the HSE guidance to determine additional hazardous factors of the project (i.e., hazardous factors of the project that are not included in the HSE guidance), then find out the to-be-controlled hazardous factors and main risks of the project through risk assessment, and develop risk control measures against additional hazards. As to the control of main risks, given that the main risks are mostly discipline-related conventional risks whose control measures have been included in the position-specific HSE guidance, there is no need to develop risk control measures against additional hazards in the HSE plan. But if the main risks of the project are unconventional risks, control measures must be included in the HSE plan.

- "Emergency response plan" is not the main content of the HSE plan but an annex to the plan for educational purpose. According to the current situation of emergency management work, at present every grassroots organization will develop a variety of emergency response plans based on the characteristics of the discipline, so there is no need to repeat the preparation of emergency response plan when compiling the HSE plan. But if the emergency response plan is not operable, you need to modify and improve it. In addition, it should be noted that the "emergency response plan" in the HSE plan and the "emergency response procedures" in the HSE guidance are different but related. The "emergency response procedures" is a part of the "emergency response plan." Being included in the position-specific HSE guidance, the "emergency response procedures" is provided for educational purpose. The "emergency response plan" is set up considering significant risks of a project, which is provided for communication before the start of a project. Once there are significant risks spreading out of control, the "emergency response plan" shall be launched immediately and call for the professional rescue force to minimize the consequences of the accident.

In order to further simplify the preparation of HSE plan for small projects or activities and to enhance the dynamic risk management of long-cycle projects, we added the risk management sheet (**Table 2**) to the version II template of the HSE plan. When using the HSE-TDOC, grassroots organizations may conduct risk management referring to the risk management sheet in the following cases:

Case one: for operating projects with long cycle and relatively fixed location (e.g., drilling of exploratory wells and critical wells and refinery shutdown for maintenance), compile the project-specific HSE plan, and incorporate the risk management sheet into the HSE plan before construction. During the construction process, carry out identification of hazardous factors on a regular basis, identify additional hazards which may arise as time changes, develop appropriate risk mitigation and control measures based on the HSE plan, and fill in the risk management sheet as a supplement to the HSE plan.

Case two: for operations with long cycle and mobile location (e.g., geophysical exploration operation and pipeline construction), compile the project-specific HSE plan, and incorporate the risk management sheet into the HSE plan before construction. During the construction process, timely identify additional hazards which may arise as time and environment change; develop appropriate risk mitigation and control measures based on the HSE plan, and fill in the risk management sheet as a supplement to the HSE plan.

Case three: for operational activities with short cycle and mobile location that are carried out in the same block (e.g., drilling of development shallow wells, downhole repair and fracturing, mud logging, wireline logging, and cementing operations that are carried out in the same block), compile the block-specific HSE plan, and incorporate the risk management sheet into the HSE plan before construction. Before single-well construction in the same block, identify additional hazards which may arise as time and environment change; and develop appropriate mitigation and control measures based on the HSE plan, and fill in the risk management sheet as a supplement to the HSE plan.

Case four: for operational activities with short cycle and relatively fixed location (e.g., production auxiliary operations, refinery temporary inspection, and maintenance), carry out hazard identification activities and fill in the risk management sheet before operation.

4.2.2.2 The compilation and application of HSE plan

The compilation of the HSE plan should be led by the major principals (team leader and project manager) of grassroots organizations. First, before the start of a project,

Code		No.		
Location (well number, job number)				
Name of the HSE plan				
1	Identification of additional hazards (including descriptions of changes in personnel, environment, process, technology, equipment, and facilities)			
2	Main risks warning (including main risks mentioned in the HSE guidance)			
3	Risk mitigation and control measures			
4	Emergency treatment			
Compiled by	MM/DD/YY		Project Supervisor	MM/DD/YY
Audited by	MM/DD/YY		Project Manager	MM/DD/YY
Relevant personnel notification records				
S/N	Name	Post (title)	Signature	Date
				Date
				Date
				Date
Completed on	MM/DD/YY		Accepted by	MM/DD/YY
Remarks:				
1. This table is an attachment to the HSE plan				
2. This table shall be filled in according to the requirements specified in the HSE plan				
3. An attached page may be added to this table if necessary				

Table 2.
Risk management sheet (sample).

relevant personnel are organized to carry out site survey and data collection to identify hazardous factors that are to be managed but not included in the HSE guidance through risk assessment, and then develop appropriate measures. The compilation of the HSE plan should be completed jointly by technicians, squad leader, key position staff, and safety officers. The finished HSE plan should be submitted to the appropriate competent department for approval according to the project risk severity. Then the approved HSE plan should be communicated to all the stakeholders and the employees taking part in the project before the commencement of the project, so as to have all the personnel involved in the project understand the project's potential unconventional risks and characteristics thereof as well as appropriate control measures. Since the HSE plan is relatively short in content, it is practical for grassroots organizations to complete the compilation and education process before the project starts.

4.2.3 The HSE checklist

4.2.3.1 Content of HSE checklist

Compilation of the HSE checklist should be based on the principle of territorial management. The territorial scope of different positions should be divided, and the territorial management categories of tools, machines, equipment, and facilities should be defined. The key components, critical parts, and vulnerable parts should be highlighted according to the relevant inspection standards.

Compared with HSE guide and HSE plan, HSE checklist is quite simple. Just like the HSE guide, HSE checklist is prepared according to different work posts. The worker should be responsible for the condition of hardware (machinery, equipment, tools) he/she uses or manages; the items need to be checked will be listed in his/her HSE checklist in order to facilitate his/her inspection; special attention should be given to the critical parts or vulnerable components of facilities. It is a form designed according to a scientifically reasonable route (order) to prompt inspection personnel to do his/her check efficiently.

4.2.3.2 Compilation and application of HSE checklist

For worksites of different natures, the ways to compile the HSE checklist are different. For example, the checklist for the standardized and customized worksites of drilling and other operations can be compiled together with the HSE guidance due to the relative fixed placement of equipment and facilities and remains relatively fixed. Different checklists should be developed according to the placement of field equipment and facilities for construction and other worksites where the placement of equipment and facilities is not fixed.

The HSE checklist is a form designed to cover all the abovementioned inspection contents according to a scientifically reasonable route (order). During shift changes or patrol inspections, relevant personnel may pay more attention to the equipment, facilities, tools, and machines under their control and be referring to the checklist, especially the critical parts of the equipment and facilities, so as to improve the efficiency of discovering hazards and ensure that the hardware facilities are in a safe condition. Although the HSE checklist is relatively simple compared with "two documents," it focuses on the inspection on the safety state of objects, which is not included in the HSE case.

5. HSE-TDOC works as project risk management mode

In the HSE-TDOC risk management mode, the HSE guidance can be used to control conventional operational risks; the HSE plan can be used to prevent unconventional operational risks, i.e., the "two documents" are to regulate human behavior; and the HSE checklist, i.e., "one checklist," can be used to inspect the state of objects. HSE-TDOC not only can be used for the risk control of mobile projects but also for the safety management of fixed workplaces. Besides, HSE-TDOC not only can be used for safety management in normal conditions but also for emergency response. Therefore, HSE-TDOC could serve as a HSE risk management mode to manage HSE risks arising from daily production and operation activities in grassroots organizations. Till now, HSE-TDOC as project risk management mode has been successfully applied in the frontline organizations of China National Petroleum Corporation (CNPC) for more than 20 years [11].

5.1 HSE guidance for conventional operational risk management

Conventional risks are the risks arising from conventional operational activities. As the content and environment of conventional operational activities are relatively fixed, the pre-established operating procedures and technical specifications such as standard operating procedures (SOP) are usually adopted for this kind of risk management. To give play to the role of risk prevention of such operating procedures and technical specifications, education for employees is required. Only through education, behavior of employees can be regulated. In order to have employees to grasp the position-specific operating procedures, the pre-established operating procedures and technical specifications are put into the HSE guidance. Through educating with the HSE guidance in daily training, the knowledge including the operating procedures can be grasped by employees, and the operating procedures can only be obeyed by them on the basis of their knowing, otherwise if they know little of the operating procedures, let alone obey them.

At present, although many enterprises have paid great attention to staff education/training, the training effect is not satisficing due to single method, disorganized contents, and lack of continuity [12–16]. To a certain extent, the HSE guidance has solved problems relating to staff technical skill training. As the HSE guidance contains all the information that should be grasped by employees such as position-specific operating procedures and emergency response procedures, it is a collection of technical skills that should be grasped by employees for a specific position. Through continuous education of the HSE guidance, employees will grasp all the knowledge required by the position and hence improve their skills to prevent risks.

5.2 HSE plan for unconventional operational risk management

The HSE plan is a document developed to prevent unconventional risks of a project that are not included in the HSE guidance. Unconventional risks not only include risks arising from a variety of unconventional activities which cannot be normalized by the pre-established operating procedures and technical specifications due to the changes such as the content or environment of the operational activities but also risks arising from various changes. It should be clear that if the risk has been already managed by either PTW (the risks arising from excavation, hot work, work at height, temporary electricity, entry into confined space, etc.) or MOC or other management tools, there is no need to mention in the HSE plan. The HSE plan is designed to manage the risks that have not been controlled by either HSE guidance or PWT or MOC or other management tools that have not been managed yet due to various reasons.

Due to the reason that the management of unconventional risks of a project is not included in the HSE guidance which is about the management of conventional risks, they can be basically managed by means of the combined use of HSE plan, PTW, and MOC. Therefore, all the identified risks (conventional risks + unconventional risks) can be managed through the combined use of HSE guidance and HSE plan.

5.3 HSE checklist for management of objects (equipment, facilities, etc.)

Based on the principle of territorial management, HSE checklist is a form designed according to a scientifically reasonable route (order) to prompt inspection personnel to pay more attention to critical parts or vulnerable components of field hardware facilities that are used or managed by employees on each position, such as tools and machines, equipment, etc., in order to improve the efficiency of discovering hidden dangers. Each position is provided with a HSE checklist. Through the combined use of the HSE checklists of all positions, full inspection on all the hardware equipment, facilities, tools, and devices can be achieved.

Based on the characteristics of each position, it is necessary to conduct an inspection on hardware equipment and facilities used or managed by the position before shifts or during working hours (whichever is applicable), to ensure that the hardware facilities are in a safe condition. Through the use of position-specific HSE checklist, unsafe state of objects will be fully inspected and effectively controlled, thereby improving the efficiency of safety check.

6. The discussion

According to the above analysis, the key problem of HSE case is it has too much content, which makes both its compilation and implementation too difficult to carry out. The main objective is to reduce its content and to make the document of HSE case a bit simple in order to play its role effectively.

On basis of the fact that the content of the HSE case is too much, we adapt the HSE case for the two documents, i.e., the HSE guidance and the HSE plan; the main problem to restrict HSE case to play its role is solved effectively. As the HSE guidance is to manage conventional risks, which has much content but is relatively stable, therefore, the HSE guidance is no deadline for completion and can be used for a long time once completed. On the contrary, the HSE plan is to manage changeable unconventional risks, for each project may have different unconventional risks; therefore, the HSE plan should be developed for each project; fortunately, its content is not so much as the conventional risks, the HSE plan is quite simple, and it can be compiled in a short time just before the start of a project [7, 17].

In this way, the greatest problem to restrict the HSE case from playing its role is solved. We can see that unconventional risks of a project that are not included in the HSE guidance can be basically managed by means of the combined use of HSE plan, PTW, and MOC. In addition, as the HSE guidance is to manage conventional risks, all the identified risks (conventional risks + unconventional risks) can be managed through the combined use of HSE guidance and HSE plan. In addition, the checklist is designed to verify whether the condition of the workplace, such as machines, equipment, tools, and so on, is safe or not. The application of "two documents and one checklist" in first-line organizations ensures that not only workers operate according to standard procedure but also that the workplace is kept in safe condition. By eliminating the causes of accidents, namely, unsafe action of workers and unsafe condition of the workplace, the model is quite effective in accident prevention.

7. The conclusion

The accident-causing theory tells us that accidents happen either because of unsafe human acts or unsafe state of objects or their combination. In the HSE-TDOC risk management mode, "two documents" are to regulate the human acts, among which the HSE guidance is to control risks arising from conventional operational activities and the HSE plan is to control risks arising from unconventional operational activities; and "one checklist" is to check the state of objects. Therefore, the HSE-TDOC risk management mode can be implemented effectively to prevent various accidents.

What is more, the HSE-TDOC risk management mode overcomes the problems of HSE case and becomes quite simple to understand and quite easy to carry out, so it is quite effective in the HSE risk management of frontline organizations for construction projects and is welcomed by our frontline organizations.

Due to the constraints of the tutorial length, the introduction of the HSE-TDOC risk management mode for construction project mentioned here may be narrow unavoidably. A systematic and detailed version has already been published by a Germany press [18] and was also introduced in textbook [19].

Acknowledgements

First of all, I must make it clear that the "2 + 1" project was officially launched by HSE Department of China National Petroleum Cooperation (CNPC), and I have been the leader and backbone of this project from the beginning to the present. Therefore, in view of the successful implementation of the project in our company and the recognition abroad, I am pleased to acknowledge all the personnel in the HSE Department of CNPC and other relevant persons, especially those who participated in the project; without their support and participation, there will be no such project, let alone its success.

Author details

Yueting Hu
QHSE Department of China National Petroleum Corporation, Beijing, China

*Address all correspondence to: huyueting@petrochina.com.cn

References

[1] International Association of Oil and Gas Producers (IOGP). Guideline for the Development and Application of Health, Safety and Environment Management Systems. Report No. 6.36/210; 2008

[2] Gower-Jones AD et al. Applied of Hazard and Effects Management Tools and Links to the HSE Case. SPE 36031; 1996

[3] International Association of Drilling Contractor (IADC). HSE Case Guidelines for Land Drilling Units; 2007. Available from: http://www.iadc.org/hsecase.html

[4] International Association of Drilling Contractor (IADC). HSE Case Guidelines for Mobile Offshore Drilling Units. Issue 3.4; 2011

[5] Shell Canada Limited. Safety Plan; 2004. Available from: http://www.mackenziegasproject.com/MGP_Nig_DPA_Section_11.pdf

[6] Hu Y et al. The principle and application of the model of "two documents & one checklist". In: Proceedings of 2010 International Colloquium on Safety Science and Technology. 2010

[7] Hu Y et al. The improvement of the model of "two documents & one checklist". In: Proceedings of the 3rd World Conference on Safety of Oil and Gas Industry. 2010

[8] Acfield AP et al. Integrating safety management through the bowtie concept—A move away from the safety case focus. In: Proc. of the Australian Safety System Conference. 2012

[9] The HSE Department of CNPC. Risk Assessment in Oil Industry (in Chinese). Beijing: Petroleum Industry Press; 2001

[10] Hu Y et al. The exploration and practice on the management of different kinds of HSE risks. International Review of Management and Business Research. 2014;**3**(4, Part 1)

[11] The HSE Department of CNPC. The Guideline to the Two Documents & One Checklist (in Chinese). Beijing: Petroleum Industry Press; 2001

[12] Serrano M, Foo J. Preventing major process safety accident events. In: International Petroleum Technology Conference, Kuala Lumpur Malaysia. IPTC, 12830. 2008

[13] Ridley J, Channing J. Safety at Work. 7th ed. U.K.: The Elsevier Ltd; 2008

[14] Reason J. Human Error. U.K.: Cambridge University Press; 1990

[15] Parker DJ. Nobody Gets Hurt. SPE/IADC Paper No. 87105; 2004

[16] The Hon Lord Cullen. The Public Inquiry into the Piper Alpha Disaster. HMSO, London, Cm 1310; 1990

[17] Hu Y. The adaptation and improvement on the mode of HSE case. In: Paper presented at the 2013 API Conference and Exhibition; Beijing, China. 2013

[18] Hu Y et al. '2+1' Programme: An Efficient HSE Risk Management Mode for Frontline Workers. Saarbrucken: Lambert Academic Publishing; 2016

[19] Li W et al. The HSE Risk Management of Petroleum Engineering (in Chinese). Beijing: Petroleum Industry Press; 2008

Chapter 3

Different Market Methods for Transferring Financial Risks in Construction

Patrick L. Brockett, Linda L. Golden and John Betak

Abstract

A goal of risk management in construction is to minimize risk exposure and the total cost of risk for a project. To this end, there are a variety of market mechanisms available for transferring risk and/or the financial consequences of a risk realization (e.g., transfer the financial consequences of a risk to an insurance company or use contractual non-insurance risk transfers such as hold harmless agreements to allocate financial responsibility to another party). Unique characteristics of construction risks are examined along with a discussion of which of these risks are insurable and which are not. The advisable risk handling mechanism to use (insurance, non-insurance transfer, retention or self-insurance, or some other technique) is provided Both the construction firm and its client must anticipate potential undesirable event occurrence with initial project planning, and build both downside risk protection and resilience into its risk management strategy. Future emerging technological advances and their impact on construction risks are discussed.

Keywords: insurance risk transfer, liability, contractual risk transfer, construction financial risk, future evolving construction risks

1. Introduction

The risk management market provides many opportunities for mitigating financial risks in construction. The risk management process consists of identifying risks, measuring risks and then deciding how to handle the risks. Once identified, risks can be avoided, retained or transferred (The A-R-T of Risk Management). There are ways of doing this, such as retaining, mitigating the risk through actions that reduce the frequency and/or severity of the risk consequence, or contractually transferring the risk to another party, either through insurance or contractual risk transfer agreements.

This chapter focuses primarily on transferring the economic (financial) consequences of losses that result from risk realization in the construction industry. We particularly explore available optimal financial risk transfer techniques, including various insurance products, and methods for transferring the financial consequences of risk realization through contractual agreements. We conclude with a section on indemnifying the financial considerations associated with new and evolving risks such as changing technology.

Construction contracts are often written with incentive clauses based on the contracted for completion date. When construction is finished ahead of schedule the contractor is rewarded a pre-specified amount per day. If the project finishes after the deadline, a pre-specified penalty is assessed for each day late. Thus, risk realization in the construction process can have twofold financial consequences: direct and indirect costs of liability and damages. We cover direct losses to property, liability to contractors, business interruption coverage (e.g., delay in start-up or completion insurance and contingent business interruption in supply chain management), worker's compensation liability, and other important insurance mechanisms pertinent to construction risk management.

2. Transference or retention of the financial consequence of risks in the construction industry

2.1 Mechanisms for risk transfer in construction

Construction firms are subject to a variety of risks with sometimes almost limitless financial consequences. Left unhandled or uncontrolled, the financial consequences of an adverse risk realization can be bankruptcy. There are several different mechanism available to the contractor (and subcontractor) which can transfer these financial consequences to another party. Contractually transferring the financial risk consequences to an insurance company by buying insurance policies designed for the specific risks affords a common method of risk transfer. A non-insurance risk transfer mechanism inserts risk transfer language into the contract of work between the contractor and other entities on the worksite so they bear the risk instead of the contractor. Each of these is discussed in more detail subsequently, along with self-insurance alternatives.

2.2 Self-insurance as an alternative risk handling technique

Not all risks can be transferred, either through insurance or through contract. According to [1], the top five uninsurable risks faced by the construction company (and needing self-insurance and risk mitigation strategies to address) are reputational risk, regulatory risk, trade-secret-intellectual property risk, political risk, and pandemic risk. With such risks the contractor must choose to either avoid the risk altogether (e.g., not bid on a contract that is deemed too risky or for which the experienced and skilled subcontractors are not available) or the contractor must retain the risk and any financial consequences internally. Alternatively, a large construction company may find risk transfer an ineffective way of hedging a particular risk, and hence choose to assume that risk; otherwise known as the self-insurance option. It is called self-insurance because it is risk financing, like insurance, but with the financial consequences paid by the company itself instead of the insurer paying. In spite of what the name may imply, self-insurance involves no transfer of risk.

All companies engage in self-insurance. Since insurance products generally have a deductible or co-pay, and a limit of liability, the contractor always faces the assumption of some of the risk (that below the deductible and above the policy limits, for example), so they are "self-insuring" these losses. Additionally, there are some risks, such as the risk of incurring criminal fines and penalties, that are not insurable, nor is there a contractual risk transfer option available. For these risks, the contractor must retain the financial consequences internally.

Self-insurance can be planned or unplanned retention. Unplanned retention occurs when the company failed to recognize a particular risk, and therefore has not prepared for addressing its financial consequences, and must pay losses internally. This can have significant consequences if losses are severe enough.

Two formal techniques for planned self-insurance are prefunding a risk account to pay for claims internally as they arise, and forming an insurance company as a subsidiary of the construction company and then buying insurance from this insurer. This insurance subsidiary is a "captive insurer". Not all companies are large enough to take advantage of these techniques, however.

Insurance companies can accept risk from others because the statistical law of large numbers and central limit theorem allow them better estimate expected losses for a risk pool, and with greater precision, than could an individual insured. By pooling a large number of similar exposures, the insurer both diversifies the assumed risk, and increases precision in estimating average losses, the basis of a premium. Administrative expenses and profit loading are added to the expected loss to arrive at a final premium to charge the insured (see [2, 3]). By knowing the expected loss for an individual insured and how much variability there is across different insureds, the insurer determines how much money they need to keep in a reserve account to pay claims with high probability.

If a non-insurance company has a sufficient number of exposure units, they can avail themselves of this same process as the insurer described above and determine the amount needed in a bank account to have sufficient funds to pay claims. The benefit of this formal self-insurance arrangement is that there is no administrative fee or profit loading charge, thus making the pre-funded bank account approach to self-insurance more economical for the company. The process may also allow for wider coverage than available on the open insurance market. Usually a company will hire a third-party administrator to assist with claims adjusting and claim payments.

The second self-insurance alternative available is to form a subsidiary that is an insurance company, and then have that insurance company write the insurance for the parent company. This subsidiary is a captive insurance company. A pure captive is an insurance company subsidiary that only insures the risks of the parent company. A pure captive is a very formal type of self-insurance since the financial consequences of the risk have not been shifted outside the original parent company. Other types of captive insurance companies can write the business of the parent as well as outside unrelated businesses. There are tax implications concerning the deductibility of the premiums paid to a captive insurer (depending on how spread the risk is between insureds), and expert tax advice is needed here. The benefit, of course, is that the profit from the insurance business is retained internally while still satisfying insurance requirements (such as the mandate to insure workers' compensation risk).

As with self-insurance generally, only very large companies can feasibly handle risk by forming a captive insurer (due to capitalization requirements). Risks in the construction industry often sent to captive insurers include workers' compensation, commercial automobile, builders risk and general liability. The captive then writes insurance policies covering these risks of the parent company.

Industry groups can also jointly form group captive insurers, and there are several in the construction industry. The benefit of joining a group captive is the additional diversification, the deductibility of premiums, and the fact that by joining an existing industry group captive, there is specialized industry expertise concerning the types of risk faced. The captive also has access to the reinsurance market (which an individual construction company does not have) and can often get insurance coverage at a lower rate than from a regular insurance company.

3. Insurance contracts facilitating risk transfer

The primary technique for transferring the financial impact of construction risks to others is through the purchase of various types of insurance. This section considers which types of construction risks are amenable to insurance and the types that are not. We then examine various important construction risks and insurance solutions to the transfer of their financial consequences.

3.1 What constitutes insurable construction risk?

Since only some risks are amenable to an insurance transfer solution, we first consider the unique characteristics of construction risk, and then describe the ideal characteristics for a construction risk to be insurable.

3.1.1 Unique aspects of construction risk

While construction is a form of manufacturing business (taking raw input materials, capital and labor to create a finished product), the differences between traditional manufacturing risk management and construction risk management are many. Risk management of construction projects is especially challenging and complex due to the unique characteristic that each project brings with it. First, the location of the construction enterprise is not fixed, as there may be several construction projects going on simultaneously resulting in many employees in various worksites and transiting between different workplaces.

The safety and risk management of each worksite must be evaluated separately (and continuously) as environmental hazards or exposures can differ from site to site (e.g., one site may have flood risk, another fire risk, another vandalism and theft risk, etc.). In international construction firms, liability risk can differ according to country and legal system. The same risk management or insurance plan will not be applicable to all projects due to location differences, beginning state and ending state site differences, differing neighboring buildings and their vulnerability, differing owners, deliverables, and contracting agreements between the owner and contractor.

Each project is also unique in terms of people working at the site. Numerous subcontractors are generally involved on a construction project, all working simultaneously at the same worksite, each subcontractor with their own contract workers, and with varying skill levels and risk culture. Coordination problems regarding safety and attitude toward risk-taking can occur. Additionally, many subcontractors are small and potentially undercapitalized, so that even if they sign a hold harmless agreement, they may not be able to live up to the assumed financial responsibility agreement (leaving no effective way to enforce it).

Depending on the terms of the contract between owner and contractor, construction projects can become adversarial due to financial pressures and uncertainties. Adversarial relationships may produce negative consequences for cooperation, safety, and the management of other risks. Fixed price contracts can exacerbate owner-contractor conflicts resulting in potential increased losses due to decreased attention to safety and risk management by the contractor (because of financial constrictions). Cost plus pricing can reduce the potential for safety and risk management related losses but increases costs. Many of these issues are also unique to construction contracts [4].

Additionally, construction projects are very labor intensive and often are performed under harsh conditions, adding to the riskiness of contracting. Management of risk becomes more important for construction since clients, specification, and workers differ from project to project.

3.1.2 Ideal conditions for insurability of a risk

Risks can be dichotomized into pure risks and speculative risks. A pure risk has a chance of loss or no loss, but no chance of a gain (e.g., a motor vehicle or a construction workplace accident). There is no gain in this situation. Speculative risks, such as investment in the stock market or contracting to build a project in the hopes high profitability, either can result in losses or gains. Pure risks are potentially amenable insurable but speculative risk are not.

However, not even all pure risks are insurable. The ideal characteristics of an insurable risk, as delineated by most risk management texts (e.g., [2]) are:

1. There should be a number of independent similar exposure units as viewed from the perspective of the insurer. This allows access to the law of large numbers from statistics to set premiums.

2. The losses that occur should be accidental or by chance.

3. A catastrophic loss should not be possible. Quite simply, a catastrophic loss, if transferred to the insurance company, could bankrupt the insurer, a likelihood not desired by the insurer. Also, catastrophes tend to violate condition 1 since adjacent properties are more likely to simultaneously experience losses making losses not independent.

4. Losses should be definite in time and measurable in loss size. Since insurance contracts are for a specified period, the insurer must be able to tell if the loss occurred during the period, and they must be able to measure the loss for claims payment and to determine premiums.

5. The probability distribution of losses should be determinable. Premium setting is essentially a statistical exercise so one must know the possible loss sizes and the likelihood of losses of various sizes to set premiums.

6. The cost of coverage should be economically feasible to provide and to buy. If the premium is unaffordable to the insured, or if the cost of underwriting (selecting and pricing) the risk is too high for the insurer, then an insurance contract will not be created.

Many risks found in construction are insurable (and discussed below). These include: workers' compensation for workplace injuries; builders risk insurance for damages during construction; general liability insurance; professional liability insurance; delay in completion insurance; insurance covering certain operational risks (such as defective construction or faulty workmanship claims); supply chain risk losses due to interruptions or damages at a supplier upon whom the contractor is depends for their own performance, and other risks like subcontractor default or financial failure.

3.2 Construction risks amenable to insurance risk transfer and relevant available insurance policies

Several standardized insurance risk transfer policies are available for use in alleviating the financial consequences of risk realization at construction sites. These policies cover different aspects of construction risk and generally satisfy the ideal characteristic of an insurable risk discussed previously.

3.2.1 Builders risk insurance

Builders risk insurance (aka "course of construction insurance") is a policy to protect the risks to property associated with a project under construction. It is insurance often written on an "all risk" basis, meaning it covers all risks except those specifically excluded by contract language. Such a policy does include a wide range of pertinent construction risk exposures such as materials, equipment, and partially completed work (completed operations however is covered under the Commercial General Liability policy). Losses can be the result of theft, fire, explosions, wind damage (except in some coastal areas), hail, glass breakage, etc. Usually excluded are ordinary wear and tear, corrosion and rust, mechanical breakdowns, employee theft, acts of war and terrorism, and damage due to faulty workmanship, materials, or planning. Builders risk insurance is essential, and covers exposures not covered under standard property risk policies since there is much higher risk of loss during the construction phase.

There is no "standard" builders risk insurance policy in the marketplace (all projects differ), so the builders risk contract should be read carefully. If the policy selected is written on an "all risk" basis it may be that certain construction defects or even faulty workmanship are covered, however this will generally depend on the contract language. Some policies have a faulty workmanship exclusion, for example. Builders risk insurance is typically project-by-project with coverage starting once the building materials are delivered to the worksite and stopping when work is complete and the finished project delivered. If a contractor or owner is going to insure several projects at the same time, they can obtain coverage on a blanket basis, which may reduce costs. Defects discovered after job completion will not necessarily be covered by builders risk insurance, and another type of insurance is needed to cover these [4].

3.2.2 Workers' compensation insurance

A very large percentage of a contractor's expenses are attributable to workers' compensation (WC) costs. Among all occupations in the USA in 2017, construction labor workers ranked as the ninth highest in terms of the number of workplace injuries and illnesses [5], and contributed 2.6% of all workplace injuries and illnesses in the USA. A 2010 report from the Bureau of Labor Statistics (BLS), said the average employer cost for workers' compensation insurance nationally was 1.6% of spending but for the construction industry this rate was 2.75 times higher (at 4.4%) [6]. A study by the National Institute of Occupational Safety and Health (NIOSH) has documented that construction industry workers experience higher rate of fatalities and injuries and higher amounts of lost work, increased WC claims and disability than the other industries. Additionally, smaller construction firms are worse, with firms having less than 10 employees being responsible for half the fatal injuries while only comprising a fourth of the construction industry [7]

All USA states (and most countries in Europe) have workers' compensation laws, and purchase of workers' compensation insurance to fulfill the statutory requirements of the WC laws is required in all USA states except Texas.

The objective of the WC system is to provide a mechanism to compensate workers' workplace injuries. The WC laws in various jurisdictions require employers to pay workers a statutory amount for work-related injuries and illnesses without regard to who caused the injury or illness, that is, the employer has strict liability (no negligence is needed for compensation). Strict liability adds additional financial incentive for employers to improve work conditions. As a counterbalancing to the WC laws, the workers' compensation system provides WC settlement as the exclusive remedy for the worker to recover damages. This means they cannot use the legal system as a remedy for costs or damages that reduces costs to the employer [2].

WC insurance provides four main coverages: medical costs for the injured worker, a reimbursement of a portion of the injured worker's wages, rehabilitation services for the worker, and death benefits of the worker who died in a workplace accident. All WC systems provide these four benefits, however the level of the each of these benefits can vary substantially state to state.

Of course, the likelihood and severity of a job injury differs significantly by employment duties, i.e., an office worker will have a much lower workers' compensation insurance rate than a carpenter or a roofer working for the same contractor. Insurers set premiums for the construction firm in accordance with the number of workers they have in each job classification [2, 4].

Several types of WC rating plans are available for larger sized insured. These include having experience rating where an "experience modifier" is created for the firm according to how their historic loss experience has been relative to the average insured's loss history. For example, if the loss history of a particular contractor is only 85% of the average contractor's loss history, then the modifier of 0.85 is applied and the premiums paid by this contractor will only be 0.85% of the manual (average) WC rate. The multiplier can also be above 1.0 if the contractor has worse than average loss experience. Experience rating provides another incentive for workplace safety to save on mandated premiums [4].

A common rating plan used by large contractors is the "retrospective rating" plan. This is similar to experience rating except the actual rate paid is determined at the end of the policy period based on actual experienced losses during the year. This retrospective adjustment of premiums at the end of the policy period can save money for doing a good job of controlling losses during the policy period. Of course, the contractor who does not control losses may be forced retroactively to pay additional premiums. Again, this provides incentives for safety and loss control. Another distinction between experience rating and retrospective rating is that in retrospective rating the contractor does not know what their premiums will be until the end of the premium period.

In construction, it is common for subcontractors on a jobsite to have their own WC insurance. A general contractor should make sure all subcontractors have WC insurance since this may affect some of the contractor's own defenses against claims by injured workers. For example, in many jurisdictions, "statutory employer immunity" that protects the owner or general contractor against claims by subcontractor's employees only applies if the general contractor has a written requirement that all subcontractors carry sufficient WC insurance [8]. For a detailed description of WC coverage, details on the history, current issues and controversies see [2].

3.2.3 Commercial general liability (CGL) insurance

A major category of insurance coverage for owners and contractors is Commercial General Liability (CGL) insurance. This generic product covers all liability exposures except those that are specifically excluded. Typical exclusions include automobile liability, workers' compensation liability, professional liability, certain injuries incurred during the construction itself, certain liabilities for faulty workmanship, and liability for completed products. Some of these can be added back by attaching an endorsement to the CGL, and most others are excluded because they are handled best by a separate policy (e.g., a commercial automobile policy, a workers' compensation policy, etc.).

The CGL policy has three major coverages: Coverage A—Bodily Injury and Property Damage Liability, Coverage B—Personal and Advertising Injury Liability, and Coverage C—Medical Payments. We examine these in turn.

In the bodily injury and property damage section, the CGL covers bodily injury or property damage caused by "an occurrence" for which the insured is legally responsible. For coverage to apply, the damage must arise from the insured's products, or completed works, or operations performed on or off site. If a lawsuit occurs, the CGL policy provides a lawyer to defend the claim.

The personal and advertising injury liability coverage (Coverage B) differs from the Coverage A in that the Coverage A is very broad whereas Coverage B only covers claims for specific offenses. If a claim does not arise from one of the listed causes, it is not covered. Another difference is that Coverage A covers damage from an occurrence resulting from negligence of the insured, which is unintentional. Coverage B, on the other hand, covers specific intentional or deliberate acts that result in harm and which arose out of business operation.

The medical payments Coverage C will pay (without their needing to be a lawsuit) for a third party's medical expenses associated with an injury from an accident occurring in the course of business activities of the insured without regard to who was at fault, and without a lawsuit. This differs from Coverage A and B where the insured needed to be responsible for the injury to be covered.

3.2.4 Professional liability insurance

Professional liability (also called errors and omissions) insurance protects a professional service provider from being held fiscally responsible in a professional negligence lawsuit. The coverage pays for defending against the claim that the insured failed to perform their professional service, or produced a professional product that did not meet normal professional standards, and that this failure to give adequate professional service resulted in a loss to the client. The coverage focuses on financial loss caused by alleged errors in professional judgment, or omissions of required and usual professional responsibilities, failure in professional oversights, or professional negligence in the service or product sold by the insured. Professional liability claims are not generally covered by a CGL insurance policy. The professional liability insurance policy is usually written on a "claims-made" basis, meaning that claims are only covered if they are made during the policy period. Common exclusions in professional liability policies are intentional or dishonest acts, and bodily injury and physical damage claims (as these are covered by CGL policies).

On the construction site, engineers, architects, electricians, plumbers, and other professionally licensed workers are held to have up-to-date professional knowledge and ability and work to professional standards. They can be held liable if their work is not up to standard and causes losses. For example, there are now professional liability lawsuits against the structural engineers, architects, and developer in the sinking and tilting 58 story Millennium Tower completed in 2009 in San Francisco, California. Because of this tilting and sinking, the tower has a minimum $200 million in repair costs, plus lost property value [9, 10].

A relatively recent product in the professional liability insurance marketplace (Contractors Professional Liability Insurance developed in the 1990s) protects contractors who engage in design-build work. Like builders risk insurance, it can be project-specific if the contractor is only doing design-build on some projects. Prior to the availability of contractors' professional liability insurance, the coverage alternative available was to add an endorsement to a design professional liability policy, and a few insurers only offered this. Coverage extended by this endorsement was typically limited to the contractor's vicarious liability for design errors and omissions inherited from a third party (e.g., an architect or structural engineer hired by the contractor), and not that of the contractor [11].

3.2.5 Commercial umbrella insurance contracts and excess liability policies

An individual primary insurance contract covers pre-specified financial consequence of a risk realization (stated in the contract) from above the specified deductible up to policy limits. If the experienced loss goes above that policy limit the contractor (or owner) is still liable for the risk consequences. Until this point in the chapter, we discussed individual primary insurance contracts like WC insurance, builders risk insurance, CGL insurance, and other primary insurance contracts (and clauses). These are viewed separately according to the risks they cover. To cover the risk of loss above the policy limits of a given liability policy, the contractor has the option of buying an additional (supplemental) policy that takes over the indemnification obligation above the maximum limits set in the underlying policy. This second policy protects the insured from potentially catastrophic losses associated with a very large liability claim. Such secondary policies are "excess insurance policies" (as they pay losses in excess to what the primary insurance pays). When the excess policy provides the same coverage details (insured events) as the primary insurance policy, the policy is a "following form excess insurance" policy. A detailed examination and discussion of the excess and surplus insurance market is given in [12].

Another possibility to raise coverage limits for an insured exposed to multiple risks is to purchase an umbrella insurance policy. The umbrella policy, at the same time and within the same contract, provides supplemental coverage in excess of the policy limits of several distinct underlying insurance policies. Thus, the umbrella policy could cover losses in excess of the policy limits of any of builders risk insurance, workers' compensation insurance or general liability policy. Instead of buying three" following form excess" policies, a single umbrella policy provides the additional limit extension to a uniform project limit that is over all the risks and is the same excess limit for all the risks covered. The umbrella policy provisions usually set a minimum on the maximum payment limit requirement for each underlying policy it spreads above since the umbrella policy is secondary, and so the umbrella insurer wants higher limits on the underlying primary policies insurance policies so they have less to pay [2].

The market for excess and umbrella policies exists to provide the contractor with an option to raise the upper coverage amounts for all underlying policy exposures to have a consistent uniform higher limit on all. Even umbrella policies have upper limits, however, so at some point the insured must be willing to self-insure large risk consequences. The maximum coverage level the contractor sets for their umbrella can be a complex choice made in collaboration with their insurance broker. If the contractor requires subcontractors to hold high limit umbrella policies, then the contractor may hold lower limits on its own policy.

3.2.6 Delay in completion or delay in start-up insurance

As noted previously, construction contracts often have incentive clauses that provide a pay bonus (per day) for finishing the project ahead of the agreed upon completion date, and impose a penalty per day for projects completed behind schedule. Unexpected delays create unexpected losses for owners, developers, construction companies, or others with a stake in the timely project completion.

There is insurance coverage available to help transfer some of this risk to an insurer for indemnification. Called delay in completion (DIC) coverage (also known as delayed completion coverage, and sometimes delayed start-up, or delayed opening coverage, or soft costs coverage (like extra accrued real estate taxes, etc.), or advance loss of profits coverage, or loss of anticipated revenue coverage), it is similar to business interruption insurance. It is written typically as part of a builders risk policy (or a marine cargo policy wherein it covers delays due to late arrival of critical shipped materials or components to the worksite). DIC policies can vary significantly from policy to policy, but DIC policy forms require the delay in completion to be caused by direct physical damage or direct physical loss to insured property. The period of indemnity is limited to an agreed upon maximum length beginning when the business that contracted for the construction would have started operation, if not for the loss. The length of the indemnity period is the time needed to remedy the delay loss. Importantly, the coverage trigger date is only applicable for start of the delay claim if the contractor can show that they would have completed on time if not for the direct physical damage or loss to insured property. To show this, the contractor may have to hire an expert, and this may be covered by the insurance.

It is important to read the policy language because not all delays are covered by all policies. Causes of delay which may not be covered depending on the contract are delays caused by having a need to redesign or rectify discovered faults or defects, damages for breach of contract, site shutdowns due to inadequate funding, or losses due to fines and penalties causing delay [13].

3.2.7 Subcontractor default insurance

General contractors compete for dependable subcontractors, particularly when construction is expanding. However, when subcontractors fail, general contractors face a host of challenges, including project delays, costs associated with work stoppage, complexities arising from trying to replace the subcontractor and potential reputation damage. Such risks tend to increase in booming construction markets, as subcontractors may take on more work than they can handle, which can exacerbate cash flow struggles. Subcontractor default insurance can help the contractor hedge this risk. In addition to contractually requiring the subcontractor have their own insurance with the contractor listed as an additional insured, and having the subcontractor agree to a hold harmless agreement written into the master contract with the subcontractors, a subcontractor default policy can be very useful.

Subcontractor default insurance, introduced by Zurich Insurance about 25 years ago, provides a way for contractors to transfer the financial consequences of subcontractor's default or non-completion of work. Until recently, few insurers have offered the product, but the market is expected to expand, and become more available to smaller contractors [14].

Retention levels on the policy (the deductible) vary from $500,000 to several million dollars, although retention levels have been going down. The premium rate charged to transfer risk to the insurer vary according to the contractor seeking coverage and depend strongly on the individual contractor's prequalification procedures for their subcontractors, on the loss history of the contractor, and on the specific loss control mechanisms implemented. The rate for subcontractor default insurance is usually fixed for 2 or 3 years [14].

The leading historical reason for subcontractor default is financial, followed by quality. There are more defaults now because of labor shortages than anything other reason. With an insured's increase in claims, insurers may make policy changes to keep the insurance viable, such as excluding coverage for problematic trades (e.g., framing) ([14], quoting Rose Hoyle).

3.2.8 Operational risks: Insurance against defective construction or faulty workmanship claims

While a large number of liability risks are covered by the CGL policy, these relate mostly to third party fortuitous or accidental bodily injury and property damage.

Most insurers have traditionally considered claims about faulty construction or workmanship as a "business risk" for the contractor. Thus, monitoring workmanship was to be taken on as a normal part of monitoring the quality of work performed while doing business, and this was viewed as being under the control of the contractor. Insurers therefore have generally excluded such claim responsibility from coverage by appending a standard "faulty workmanship" exclusion clause to the CGL policy.

If the contractor's completed work or product is faulty, or if the work is not what was contractually specified, the contractor's unendorsed CGL policy will generally not cover the costs to remediate it (but see the builders risk section for in-progress claims). A California court elucidated this as follows, "Generally liability policies... are not designed to provide contractors...with coverage against claims their work is inferior or defective.... Rather liability coverage comes into play when the contractor's (insured) defective materials or work cause injury to property other than the insured's own work or products." See Clarendon America Ins. Co. v. General Sec. Indem. Co. of Arizona, 193 Cal. App. 4th 1311, 1325 (2011), sited in [15].

The contractor can, however now buy an endorsement covering faulty workmanship from some insurers [16, 17]. These endorsements provide funds for claims due to faulty workmanship, materials, or products, even if discovered after the project termination. It is worth noting, however, that the coverage is only applicable for policies in force, so terminating (canceling) the policy when the project is done but before the expiration of the statute of limitations for clams has expired may leave a risk exposure for late filed claims. The contractor should check coverage with a broker since coverage interpretation of the CGL language is on a state-by-state basis, and many insurers have now created new coverage endorsements redefining the scope of coverage.

3.2.9 Supply chain risks for contractors and contingent business interruption (CBI) insurance

Supply chain risk is created by disruption in the sequencing of permitting, subcontractors' arrival for work, and the arrival of materials at the worksite when needed. Additionally, particular owner specified items can also be problematic to source, and owner-imposed requirements and impacts need to be documented to help manage this risk. Demand for globally sourced products such as marble from Italy, Saltillo tile from Mexico and machinery from Germany have increased. At the same time, the supply chain inventory for these products has become "leaner" and the use of "just in time" inventory control has grown in response to a competitive desire to increase efficiency and save inventory or holding costs. When the supply chain is properly functioning, such processes can result in cost savings. On the other hand, losses can occur if suppliers have disruption, such as an earthquake in Mexico or Iceland's Eyjafjallajökull volcano that shut down air traffic over much of northern Europe in 2010 (disrupting supply chains worldwide). These natural catastrophes can cause delays in the arrival of construction material and construction progression can suffer. Since the damage did not occur to the construction project's own physical site, losses associate with these supply chain disruptions will generally not be covered by the usual builders risk, general liability, or the contractor's other policies.

There is an insurance policy that covers the risk of a supplier having damages that affect the contractor's ability to perform on their own construction project. This product is Contingent Business Interruption (CBI) Insurance. It covers losses to the contractor due to a disruption or delay in receiving products, components, or services from a supplier because of an incident at a supplier's property. Non-physical damage events affecting the supplier could include strikes, pandemics; civil or military action; and regulatory actions against the supplier. The CBI policy can be written to cover either incidents at the location of a particular single named supplier or it could cover all suppliers depending on the terms of the contract. Coverage under these policies is triggered by interruption to contractor due to supply chain or logistical failure [18].

It should be noted that Contingent Business Interruption Insurance is different from regular Business Interruption (BI) Insurance. CBI covers the risk of damage (loss) to the contractor due to an incident at a supplier's location. On the other hand, regular BI Insurance addresses the risk of losses arising at the contractor's worksite that cause losses and interruptions to the contractor.

4. Non-insurance risk transfer: contractual transfer embedded within other contracts

The contract between the owner and the general contractor (or the contractor and the subcontractors) specifies the terms and conditions, details of construction, material, deadlines for completion and many other project specific details. The contract also identifies and allocates risk. Some risks that might be borne by one party can be transferred by mutual agreement to another party in the contract. Here we consider several risk transfer mechanisms available to the two parties signing the master construction contract that can be embedded within the master contract.

4.1 Risk transfer as part of subcontractor agreements

The decision as to who bears the risk in a construction project should generally worked out contractually. Risk created by a subcontractor or its employees can still come back to affect the contractor through the legal doctrine of *respondeat superior* and the existence of vicarious liability of the contractor (the liability of an employer or supervisor for liability generated by their employees). Often contracts are written between the contractor and subcontractor in such a manner as to make sure the risks created by a subcontract do not adversely affect the contractor. There are several important techniques to transferring risks contractually, and we discuss these below.

4.1.1 The contractor as an additional on subcontractor's insurance

An insurance contract is a legal contract between the insured and the insurer that agrees to pay specific amounts for claims filed within the policy period that satisfy the terms of the policy. A liability insurance policy such as the CGL policy, for example, will pay any liability claim amount (damages) that meets the conditions of the contract plus litigation costs up to the specified policy limits. Since the policy is a contract between the insurer and the insured, only the insured can file a claim against the policy. Thus, for example, if a contractor hires a subcontractor who causes physical damage, bodily injury or liability expense related to the construction project, only the subcontractor can file a claim on their insurance policy. Since filing of claims can make subsequent experience rated insurance purchases more expensive, the subcontractor may be reluctant to file a claim. A way around this is for the contractor to have written into their general construction contract with the subcontractor that they (the contractor) be listed as an additional insured on the subcontractor's insurance policy. This gives them equal status to talk with the subcontractor's insurer, and the contractor now has the ability to file claims against the subcontractor's policy.

If there is a claim the contractor has against the subcontractor that would trigger coverage by the subcontractor's insurance policy, the contractors can give permission for their own insurer to deal directly with the subcontractor's insurer, as they are a party to both contracts. By using the additional insured route to the subcontractor's insurance policy, the contractor can have the requisite damages and defense costs paid without drawing upon the policy limits of any other policy they might have. This also saves the contractor money on experience rated insurance policies, as the adverse claim experience does not go on the contractor's claim record.

It is also desirable that the contractor have written into their contract with the subcontractor that they be listed as having primary (as opposed to excess) additional insured status on the subcontractor's policy. Primary insured status means that the subcontractor's policy becomes the primary policy (pays first) instead of the contractor's own policy when a claim is filed, and it will pay up to the policy limits of the subcontractor before tapping any of the contractor's own insurance policies. The contractor's policies are then secondary insurance and pay whatever is left on the claim above the primary insurance policy's limits. Transferring claim costs to the subcontractor's policy helps control the contractor costs and allows them to retain their own policy coverage unused. If the contractor were listed as an additional insured on an excess basis, then the contractor's own policy becomes primary (and pays first up to policy limits) and the subcontractor's policy becomes excess and only pays the costs in excess of the payment under the contractor's policy.

Many contractors write into their original agreement that they be continued as an additional insured for as long as possible since claims may arise long after the subcontractor leaves the worksite. The contractor can mandate they obtain a Certificate of Insurance from the subcontractor that shows coverages as well as listing the contractor as an additional insured.

Several different forms and endorsements exist for listing the contractor as an additional insured on the subcontractor's policy. The most favorable risk transfer (for the contractor) is to have additional insured status with an endorsement that includes both work in progress and completed work (an ongoing operations endorsement and a completed operations endorsement). These endorsements can be recommended by the contractor' insurance broker [19].

4.1.2 Owner and contractor controlled insurance programs and wrap-up insurance

Every construction project contains multiple subprojects and multiple sources of potential risk of losses. The larger the project, the more subcontractors there are on the project, the more varied, complex, and potentially overlapping are the risk and potential losses. In smaller or traditional construction projects, each subcontractor takes care of their own risks through their own insurance, and the contractor requires a hold harmless agreement and to be listed as an additional insured. With a large-scale project, (e.g., $50–100 million) there are savings by having all contractors or subs covered under a single policy. Because of the potential interactions of different subcontractors, there can be duplicative coverage for some risks, and disagreement (and litigation) among subcontractors (and their insurers) as to fault. Subcontractors have their own insurer giving the potential for litigation among insureds as to who pays first. There can also be lack of uniformity of policy limits, conditions, terms and conditions specified by each insurer. Finally, the owner should be listed as an additional insured on all relevant policies (e.g., contractor and sub-contractors), which may create costly duplicative coverage of owner's risks.

A solution to this situation is for one party to obtain insurance policies that covers multiple other parties working on the construction project. One insurance policy covers the entire project instead of each of the multitude of subcontractors each with their own insurance policy covering just their piece. This arrangement to have one insurance policy cover the entire project is a wrap-up insurance program, as all subcontractors' risks are "wrapped up" into a single policy. The goal of a wrap-up program is to reduce total insurance costs for the project while affecting consistent coverage. If the owner is the lead party who arranges for the single insurance policy that all contractors and subcontractors subscribe to, the arrangement is an Owner Controlled Insurance Program (OCIP). If the general contractor is the lead party with subcontractors as subscribers, the arrangement is a Contractor Controlled Insurance Program (CCIP). A number of large contractors are now considering wrap-up insurance programs, and CCIPs are much more common today than in the past [8].

There are several advantages of a wrap-up insurance program (either OCIP or CCIP). First, it provides uniformity of coverage with a single insurer. This eliminates duplicative coverage and differences in conditions and limits. It eliminates costly legal bickering between the subcontractors' insurers over who has responsibility of a claim, which can eat into the policy limits of the coverage. It allows for more advantageous "economies of scale" in negotiating with the insurer over price. All these factors can reduce total premiums. Subcontractors pay their "share" of the premium and do not get project insurance on their own.

Centralized loss control and safety policies can be affected by using the wrap-up plan, making for uniform loss control incentives. Importantly, the wrap-up program can complicate the bidding process as the use or non-use of the wrap-up arrangement can greatly affect each subcontractors' insurance related costs. For effective bidding, subcontractors must know their insurance costs, thus, the creation and details of the wrap-up arrangement must be explicitly determined before bidding and project commencement.

The goal of the OCIP or CCIP is to save insurance costs so it usually only includes coverages for which there would be cost savings by having the individual policies wrapped up into a single policy. Typically, these include workers' compensation, CGL, builders' risk, and sometimes umbrella insurance coverage. Other coverages like commercial automobile or professional liability do not offer the potential cost savings and are not generally included in the wrap-up program but rather continue coverage by individual subcontractors [4].

4.2 Hold harmless and indemnification agreements

A hold harmless agreement is a contractual agreement between two parties that specifies how the risk of liability arising during construction will be distributed. The contracting parties to the hold harmless contract agree among themselves, before any loss occurs, on how to split the costs of a risk realization. Usually hold harmless agreements are embedded clauses within the general construction contract and they shift the risk from one party (who originally holds the risk) to another party. From an economic efficiency perspective, this transfer might be done in order to place the financial responsibility with the party that has best control over the risk, hence creating an enhanced financial incentive to control risk by the party that best has the ability to control the risk. Alternatively, the transfer of risk might place the risk with the party that has a comparative economic advantage in risk bearing so that the cost of risk is lessened [4].

The two parties are the" indemnitor" (the one who agrees to indemnify or hold harmless) and the "indemnitee" (the one who is originally potentially liable to pay but who has transferred this risk to the indemnitor and can no longer be harmed by the financial burden). Illustrative examples include having the owner as the indemnitee and the general contractor as the indemnitor, or it could be a contractor as the indemnitee and subcontractor as indemnitor.

As an illustration of the incentive effects, an electrical subcontractor has best control over how the wiring in a construction project is performed. Faulty wiring however, could cause a financial loss for the contractor, such as if a third party was injured and sued the contractor. If the contractor had the subcontractor sign a hold harmless agreement, then the subcontractor has agreed to pay for any harm to the contractor caused by subcontractor's work (within the terms of the hold harmless agreement). The financial consequences of the risk of faulty wiring would be transferred to the party best able to ensure there is no faulty wiring. This hold harmless and indemnification clause ensures subcontractors monitor their own work, as they bear the consequences of their losses.

The type or form of hold harmless/indemnification agreement determines the degree to which the liability associated with the indemnitee's negligence is shifted to the indemnitor. There are three common forms of indemnity (hold harmless) agreements: (1) a broad form, (2) an intermediate form, and (3) a limited or comparative fault form [4, 20].

First, the broad form transfers the most incurred risk (financial responsibility) from the contractor (indemnitee) to the subcontractor (indemnitor). With this broad form agreement, the subcontractor agrees to take on all related liability for accidents whether it be due to their own negligence, negligence by the contractor, or a combination of negligence on the part of both. Due to its broad scope, the subcontractors must usually get an additional insurance policy on top of their own liability policy. Note also that since the subcontractor with this type hold harmless form has agreed to take on the contractor's liability, even that which had nothing to do with the subcontractor; there is an adverse incentive for safety created for the contractor to take care and spend money on safety in the workplace. Therefore, some jurisdictions have declared the broad form illegal.

The second intermediate type of hold harmless agreement has the subcontractor (indemnitor) assume responsibility for all loss costs except those arising solely from the contractor's (indemnitee's) negligence. This is the most common hold harmless agreement type. If both the subcontractor (indemnitor) and the contractor (indemnitee) are partially negligence the subcontractor is responsible for all liability.

The third limited form hold harmless agreement holds the subcontractor (indemnitor) responsible only for their part of the liability and the contractor (indemnitee) is responsible for his or her part. This is a comparative fault form, as determination must be made as to what percentage of the liability was the fault of the subcontractor and what was due to the contractor [20].

It should be noted that the party agreeing to assume the liability of another under a hold harmless agreement might, but does not automatically, have recourse to their CGL policy to cover their contractually assumed liability. The 2013 CGL policy has a "contractual liability exclusion" that eliminates an assumption of such risk within the liability section of the CGL unless it is for a liability that the insured would have had even without having signed a hold harmless agreement, or unless it was for a liability assumed in a contract or agreement that is an "insured contract." The meaning of this last term continues to be litigated, and it behooves the contractor to consult their broker for what parts (if any) of the hold harmless agreement can be covered by the CGL. Court rulings have differed by state [21]. Many conclude that the hold harmless agreement is an "insured contract" and hence is excluded from this policy exclusion (and therefore is included in the CGL coverage).

5. Surety bonds for construction projects

Like insurance, surety bonds exist to ensure that a construction project is completed within the contract's terms and conditions. Most surety bonds are underwritten by sub-divisions of insurers, and like insurance, surety bonds are regulated at the state level in the USA by the state's Department of Insurance. Surety bonds are not insurance, however, but rather provide a guaranty that the obligations of the contractor will be fulfilled. The Surety (the entity writing the bond) can assist the contractor if the contractor experiences cash flow problems. If the contractor fails to perform or is held in default of the contract, or abandons the project, the Surety may replace the contractor to get the project completed.

Unlike insurance, written to cover unexpected fortuitous events that affect the project and that indemnifies the insured and provides legal defense of the insured under the policy, a surety bond is written to cover the contractor's obligation to the owner under the contract and does not provide any legal defense for the contractor. An insurance contract has a specific period for coverage and is renewable whereas a surety bond is generally project specific and lasts throughout the project. If an insurer makes a payment on behalf of the contractor, the contractor is not expected to reimburse the insurer, whereas if a surety bond provider makes payments on behalf of the contractor, the contractor must pay them back. Because the underwriting of the bond involves contractor prequalification based on their construction experience and financial strength, the bond is usually underwritten with the expectation of no loss. When used in construction, surety bonds are called Contract Surety Bonds [8].

Unlike an insurance contract, which is between two parties (the insurer and insured), the surety bond involves three parties: the Obligee (project owner or contract beneficiary), the Surety (who writes the bond and promises performance of the contract), and the Principal (contractor who contracted to construct according to the contract).

Three types of Contract Surety Bonds are most relevant in construction. These are (1) the "bid bond" which protects the Obligee should the contractor be awarded the contract and then either does not sign the contract or does not provide the called-for payment or performance bonds, (2) the "payment bond" that guarantees that the contractor will pay workers, suppliers, and sub-contractors, and (3) the "performance bond" that protects the Obligee from loss should the contractor fail to perform on the construction project according to contract. A Surety assures the project is completed according to contract [8].

Surety bonds are very important for handling the financial consequences of certain risks in the construction industry since many entities require a surety bond from the contractor or sub-contractors as a condition of awarding the contract. For example, general contractors may require their subcontractors to provide surety bonds to protect the contractor. In the public sector, statutory requirements by federal, state and local governments require contractor bonding to ensure the lowest bidder can actually perform on the contract and that suppliers and subcontractors will be paid and taxpayer money be well spent. In the private sector, lending institutions may require surety bonds (and might even become a dual obligee on the surety bond) to protect their investment. Private owners, especially on large projects, may require the contractor provide a surety bond to guarantee the quality of the contractor (since they are pre-qualified as discussed previously) and to make sure their project gets accomplished according to plan in the event of contractor default of failure.

6. Emerging market technologies affecting construction risk

There are many emerging risks dues to world dynamics and risks in the market. Construction managers will likely have to respond to these in their risk management processes or pay the consequences. Some insurance providers already have products to address these. Through the use of insurance providers, such as Lloyds of London, construction managers can negotiate new insurance products that meet their specific emerging risk management needs or choose to self-insure. This section is forward facing to identify some emerging risks that demand construction management attention before the risks are devastating.

The construction industry is one of the least automated industries, relying heavily on human labor. There are, however, different types of construction robots now poised to revolutionize parts of construction. The use of construction robots can increase efficiency and decrease cost, but also can create risks and uncertainties relatively unfamiliar to construction risk management [22–24].

One potentially disruptive technology is 3D-printing that can build even large buildings on demand. A robotic arm controls a 3D-printer and this 3D printer produces an entire building (or component parts needed for construction). This technology has been used for canals and bridges, with a 3D printed canal built in Netherlands in 2014, and the first ever-3D printed pedestrian bridge built in Spain in December 2016 [24].

Robots may dramatically improve the speed and quality of construction work [22–24]. It was announced recently that Sunconomy, a USA construction company, received permits to build its first 3D printed manufactured house in Lago Vista, Texas [25]. WinSun, a Chinese construction company, expects up to a 50% savings on housing construction using 3D printing [26].

All forms of construction robots could fundamentally change risks, from risks associated with injuries, to project completion time, to supply chains [27]. However, two areas of liability exposure may arise: products liability and intellectual property violations (the 3D plans used).

Contractors using 3D printing should check their CGL policy as many have exclusions for cyber related risk and may exclude liabilities associated with embedded software errors that cause product defect loss when using 3D printing. Contractors should consider getting a version of products liability insurance to cover these losses. Insurance risk transfer issues associated with this emerging technology are discussed in [28]. Demolition robots are another robot that, while slower than demolition crews, are safer and cheaper [29] but create liability.

Emerging AI based applications can be very beneficial to construction. These include: AI innovations providing enhanced visual processing using videos of worksites to help identify safety hazards, drones, high tech sensors and other enhanced visual processing to automate tracking of project progress against plans, as well as 3D models from data captured by drones to measure progress against original designs, and to detect any errors or inconsistencies [30].

In spite of these and other benefits of AI and tech innovations, they do create liability transfer risks still not well identified or addressed. These insurance liability transfer risks are very complex and the party responsible for AI and innovation failures causing damages have yet to be legally decided [31]. Cyber liability exclusions in the CGL may cause lack of coverage issues and it is important for construction managers to recognize and deal with these risks.

7. Conclusion

There are many risks in construction necessitating decisions to avoid, retain or transfer an identified risk (The A-R-T of Risk Management) that ideally should be made in the planning phase before project start. This chapter delineated characteristics of construction risk and focused on ways to transfer financial risk to the insurance market, to other stakeholders, to retain or to avoid that part of the business creating the risk.

A contractor's goal is to minimize the cost of risk, so alternative risk transfer methods were discussed, from well-established ones to emerging ones. Builders can contractually transfer risks to involved others or clients (e.g., through hold harmless agreements) or to insurance companies. The marketplace is dynamic, and transfer options for construction risks are continually evolving.

This chapter looked forward and discussed emerging technologies that will be creating new risks to anticipate (e.g., the advent of 3D printing, robotics, and AI). Technological advancements will always present new risk challenges.

Finally, issues of sustainability (the ability to have low environmental impact) and resilience (the ability to bounce back from unexpected or catastrophic events) will become increasingly important for construction risk managers. This is partially due to climate change, increasing catastrophic events, and the consequential regulatory changes likely to spur new and challenging building codes. These are among other currently unknown and, as yet unaddressed risks are important for the construction manager to anticipate.

Author details

Patrick L. Brockett*, Linda L. Golden and John Betak
University of Texas at Austin, Austin, Texas, USA

*Address all correspondence to: utpatrickbrockett@gmail.com

References

[1] Freedman A. Unquantifiable Exposures: Top Five Uninsurable Risks. Risk & Insurance [Internet]. September 2, 2014. Available from: http://riskandinsurance.com/top-five-uninsurable-risks/ [Accessed: Jan 11, 2019]

[2] Baranoff E, Brockett P, Kahane Y, Baranoff D. Risk Management for the Enterprise and Individuals [Internet]. V2.0. Boston: Flatworld Publisher; 2019. Available from: http://www.flatworldknowledge.com/

[3] Brockett P, Cox S Jr, Witt R. Insurance versus self-insurance: A risk management perspective. The Journal of Risk and Insurance. 1986;**53**:242-257

[4] Witt R. The Optimal Allocation of Insurance Related Risk and Costs in Construction Projects. Austin: University of Texas at Austin and The Construction Industry Institute; 1993

[5] Insurance Information Institute (iii). Facts + Statistics: Workplace Safety/Workers Comp, Table 10 [Internet]. 2018. Available from: https://www.iii.org/fact-statistic/facts-statistics-workplace-safety-workers-comp [Accessed: Jan 9, 2019]

[6] WorkCompLab. Workers' Compensation Insurance by Industry. 2018. Available from: https://workcomplab.com/insurance-industry/ [Accessed: Jan 8, 2019]

[7] Schofield K, Alexander B, Gerberich S, MacLehose R. Workers' compensation loss prevention representative contact and risk of lost-time injury in construction policyholders [Internet]. Journal of Safety Research. 2017;**62**(Supplement C):101-105. DOI: 10.1016/j.jsr.2017.06.012 [Accessed: Jan 10, 2019]

[8] Muse F, Kneisel C, Robert F. Insurance coverage for construction projects: Chapter 4. In: Cushman R, editor. Construction Business Handbook. New York: ASPEN Publishers; 2003. ISBN-0735536805. Available from: https://www.kilpatricktownsend.com/~/media/Files/articles/Construction%20Business%20Handbook.ashx [Accessed: Feb 25, 2019]

[9] Tarmy J. Who will foot the bill for San Francisco's $750M millennium tower? Insurance Journal [Internet]. 2017. Available from: www.insurancejournal.com/news/west/2017/02/02/440648.htm [Accessed: Jan 3, 2019]

[10] Wertheim J. San Francisco's leaning tower of lawsuits [Internet]. CBS News, CBS Interactive. 2018. Available from: www.cbsnews.com/news/millennium-tower-san-francisco-leaning-tower-of-lawsuits-60-minutes/ [Accessed: Jan 10, 2019]

[11] International Risk Management Institute (IRMI). Contractors Professional Liability [Internet]. 2019. Available from: https://www.irmi.com/term/insurance-definitions/contractors-professional-liability-insurance [Accessed: Jan 10, 2019]

[12] Brockett P, Witt R, Aird P. An economic overview of the excess and surplus lines insurance. Journal of Insurance Regulation. 1990;**9**(2):234-258

[13] Marker S. Delay In Completion Losses Under A Builders Risk Policy. Merlin Law Group, Property Insurance Coverage Law Blog [Internet]. 2014. Available from: https://www.propertyinsurancecoveragelaw.com/2014/07/articles/commercial-insurance-claims/delay-in-completion-losses-under-a-builders-risk-policy/ [Accessed: Jan 11, 2019]

[14] Lerner M. Subcontractor default insurance market set to expand. Insurance journal [Internet]. 2018. Available from: https://www.businessinsurance.com/article/20181218/NEWS06/912325707/Subcontractor-default-insurance-market-set-to-expand [Accessed: Jan 8, 2019]

[15] Wakefield J. When Does Insurance Cover Faulty Workmanship? Construction Litigation Blog, Insurance Blog, James Wakefield, Cummins & White, LLP [Internet]. 2012. Available from: https://www.cumminsandwhite.com/when-does-insurance-cover-faulty-workmanship/ [Accessed: Jan 10, 2019]

[16] Citizens General Insurance Brokers. Does Your General Liability Insurance Cover Faulty Workmanship? [Internet]. Available from: https://citizensgeneral.com/business-insurance-news/postid/69/does-your-general-liability-insurance-cover-faulty-workmanship [Accessed: Jan 11, 2019]

[17] Paperless Insurance. Faulty Workmanship Coverage Endorsement [Internet]. 2013. Available from: https://www.paperless-insurance.com/faulty-workmanship-coverage-endorsement/ [Accessed: Jan 10, 2019]

[18] International Risk Management Institute (IRMI). Contingent Business Interruption: Getting All the Facts [Internet]. 2019. Available from: https://www.irmi.com/articles/expert-commentary/contingent-business-interruption-getting-all-the-facts [Accessed: Jan 10, 2019]

[19] DBH Resources, Inc. Contractors Risk Management Practices, an Educational Guide. 2008. Available from: http://lucienwright.com/Contractors_Risk_Management_Guide.pdf

[20] Rodriguez J. 3 Types of Hold Harmless Agreements and When to Use Them. Small Business. 2018. Available from: https://www.thebalancesmb.com/types-of-hold-harmless-agreement-and-when-to-use-844792 [Accessed: Jan 4, 2019]

[21] Stanovich C. Expert Commentary: Contractual Liability and the CGL Policy [Internet]. International Risk Management Institute (IRMI). 2018. Available from: https://www.irmi.com/articles/expert-commentary/contractual-liability-and-the-cgl-policy [Accessed: Jan 10, 2019]

[22] Robotics Online Marketing Team. Construction Robots Will Change the Industry Forever [Internet]. 2018. Available from: https://www.robotics.org/blog-article.cfm/Construction-Robots-Will-Change-the-Industry-Forever/93 [Accessed: Feb 25, 2019]

[23] IIT Madras Printability Lab. 5 Ways that 3D Printing Is Changing the Global Construction Industry [Internet]. 2018. Available from: https://imprint-iitm.org/3d-printing-changing-the-construction-industry/ [Accessed: Feb 25, 2019]

[24] Honrubia M. 3D Printing and its Application in the Construction Industry [Internet]. 2018. Available from: https://www.ennomotive.com/3d-printing-and-its-application-in-the-construction-industry/ [Accessed: Jan 6, 2019]

[25] Vialva T. Sunconomy to develop 3D printed concrete homes in Texas. 3D Printing Industry [Internet]. 2019. Available from: https://3dprintingindustry.com/news/apis-cor-and-sunconomy-to-develop-3d-printed-concrete-homes-in-texas-146575/ [Accessed: Jan 10, 2019]

[26] Burger R. The Effect of 3D Printing on Global Construction [Internet]. 2019. The Balance Small Business [Internet]. Available from: https://www.thebalancesmb.com/3d-printing-construction-industry-845342 [Accessed: Jan 25, 2019]

[27] How 3D Printing in Construction is Transforming the Industry [Internet]. 2015. Available from: https://contractorsinsurance.org/3d-printing-in-construction/ [Accessed: Jan 25, 2019]

[28] Friedman DJ, Buschmann C. What You Need to Know About 3-D Printing and Insurance [Internet]. 2017. Available from: https://www.industryweek.com/additive/what-you-need-know-about-3-d-printing-and-insurance [Accessed: Jan 25, 2019]

[29] Hooker K. Advantages of Demolition Robots. Concrete Construction [Internet]. 2015. Available from: https://www.concreteconstruction.net/how-to/repair/advantages-of-demolition-robots_o [Accessed: Jan 5, 2019]

[30] Clavero J. Applications for Artificial Intelligence in Construction Management. eSub [Internet]. 2018. Available from: Available from: https://esub.com/applications-artificial-intelligence-construction-management/ [Accessed: Feb 25, 2019]

[31] Kerr M. Artificial intelligence ties liability in knots. Risk & Insurance [Internet]. 2017. Available from: https://riskandinsurance.com/artificial-intelligence-ties-liability-knots/ [Accessed: Jan 25, 2019]

Chapter 4

Holistic View on Multi-Stakeholders' Influence on Health and Safety Risk Management in Construction Projects in Tanzania

Sarah Phoya and Krystyna Pietrzyk

Abstract

Construction projects constitute complex and dynamic systems, which pose high health and safety risks to the practitioners. As a result, many researchers have underscored the importance of risk management that requires inputs from all stakeholders across different stages of the construction project from the design up to the construction phase. However, there is a limited knowledge about stakeholders' influence on health and safety risk management in building construction projects in Tanzania. To fill this gap, a case study approach was employed to analyse three large ongoing construction projects in Dar es Salaam in Tanzania. Data were collected through questionnaire survey and in-depth interview with a range of stakeholders: clients, consultants, contractors, workers and regulatory agencies. From the findings and with reference to literature, the systems thinking approach was used to develop a model showing the stakeholders' influence on health and safety risk management. The pattern of relationships between different stakeholders and the capacity of the system to offer health and safety control was traced using the results of the case studies of the three projects. The study confirms that there is a link chain relationship when stakeholders influence the health and safety risk management at the following stages of the construction projects in Tanzania.

Keywords: health and safety, risk management, stakeholders, building construction projects, system

1. Introduction

Generally, literature on risk management ascertain that all key project stakeholders (clients, designers, sub-contractors, contractors, and statutory authorities) should be involved in considering safety systematically, stage by stage at the outset of the project [1]. In fact, many health and safety risks arise due to lack of risk management from initiation of project to the completion of construction projects [2, 3]. This underscores the fact that, health and safety risk in construction project originates upstream from the building process itself and is connected to processes such as planning, scheduling, design, tendering and construction.

In view of the redistributive impact of poor safety performance, all stakeholders involved in different stages of construction project should be accountable for safety

IntechOpen

risk management [4]. In a similar vein, Charles et al. [5] and Zhang et al. [6] also emphasised the importance of developing communication networks throughout the construction process and well-articulated responsibilities for the stakeholders involved in the project. Furthermore, in its 1992 code of practice, on 'Safety and Health in Construction', the International Labour Organisation (ILO) outlined the responsibilities of health and safety among different groups in construction project. The ILO [7] advised that national laws of different countries should include responsibilities of health and safety risk to clients and designers (engineers, architects and quantity surveyor) in construction projects.

Several studies have been conducted on project stakeholders' influence on health and safety risk management. Some of the authors have focused on safety responsibility among designers during the design phase [8–11] while others focused on the role of clients on health and safety management [12–16]. They maintain that clients have a major role in project implementation, and therefore, they should push for the safety requirements. Furthermore, Well and Hawkins [17] have focused on safety in the procurement phase. They argue that addressing the issue of construction safety in the design and procurement phase could have a substantial impact on reducing injuries and costs associated with safety-related delays in projects.

Notably, all these studies have focused on individual stakeholder and their roles with emphasis on their individual viewpoints on risk management. Consequently, there has not been much study focused on the mechanisms on how these stakeholders can co-operate as a team to influence health and safety risk management in a systems thinking model. It is not known how project stakeholders would interact, communicate, deal with risk information, let alone on their roles, liabilities and responsibilities which influence health and safety risk management. This study therefore explores the influence of multi-stakeholders such as clients, design teams and contractors on health and safety risk management in construction projects in Tanzania using systems thinking model. The aim is to develop holistic understanding of multi-stakeholders' influence on health and safety risk management.

2. Theoretical framework

2.1 The concept of stakeholders

A stakeholder is a relatively recent term coined originally for the corporate sector. Freeman [18] defined a stakeholder as a person or an entity that can affect or is affected by the accomplishment of an organisational or project purpose. Mitchell et al. [19] classified stakeholders into definitive stakeholders, expectant stakeholders and latent stakeholders based on their power, legitimacy and the urgency of their claim on the organization. Clarkson [20] classified stakeholders into primary stakeholders, on whom a corporation depends for its survival, and secondary stakeholders, as those who are not essential but have influence on or are influenced by the corporation. Both Leung and Olomolaiye [21] and Olander and Landin [22] categorise stakeholders as either internal (clients, consultants or contractors) or external (external public or external private parties) to a project. Internal stakeholders are those involved in the decision-making process, whereas external stakeholders are most often affected by the potential outcome of the project, either directly or indirectly as stakeholders. Here, stakeholders are considered as those whose performances play an important role in determining a project's success. These stakeholders include project clients, project management consultants (architects, engineers and quantity surveyors) and project contractors, sub-contractors, workers and regulators and legislators in the legal system. The strong cooperation of stakeholders is necessary for project success, since a project can be considered a temporary organisation of stakeholders pursuing an aim together.

2.2 Health and safety risk management

Risk is regarded as the measure of probability (likelihood) and consequences of not achieving the defined goal [23]. Therefore, risk event has two primary components, that is a probability/likelihood of occurrence of an event and impact of the event—amount at stake. In that regard, risk is considered as a function of likelihood and impact [24]. Risk in this research means the possibility of suffering harm or loss, a factor, a cause of element involved in certain danger and its severity in individual or enterprises in informal construction sector. The sources (hazards) of health and safety risks on construction sites are identified as: nature and physical layout of the work space, location and weather, equipment and hazardous materials, human behaviour and attitude, leadership, and safety culture of the organisation.

Risk management is defined as 'a systematic way of looking at areas of risk and consciously determining how each should be treated. It is a management tool that aims at identifying sources of risk and uncertainty, determining their impact and developing appropriate management responses' [23]. The overall goal of risk management is to maximise the opportunities and minimise the negative consequences of risk threats in a project [25]. Therefore, as a process, RM should be cyclic and dynamic in nature and is important to be established early in a project and continually addressed throughout the project lifecycle, and it should be proactive rather than reactive, involving all stakeholders in the project. Generally, risk management involves process in risk identification, risk analysis and risk response [23, 24].

2.3 The concept of systems thinking

Senge [26] describes a system as a perceived whole, whose elements belong together because they affect each other over time and operate towards a common purpose. It focuses on holistic perspective emphasising the interplay between the systems and their elements in determining their respective functions. The interaction between the system's elements can be complex with simultaneous mutual influences rather than the linear cause and effect chain relationship [27]. The elements in the system may be tightly and strongly linked and change in response to each other, therefore, indicating strong interdependence of the system's components.

Construction projects are complex systems involving multiple and mutual components. Thus, construction projects consist of many interacting stakeholders such as clients, contactors, consultants and workers with different management objectives and functions that contribute to the whole. Thus, each stakeholder in a project has specific roles to play to achieve a collective project goal. However, the roles of the stakeholders are quite interrelated and insufficient performance of one of them directly affects the project's goal achievement no matter how well other stakeholders perform their roles. To understand this type of relationship, a systems thinking approach needs to be employed. Reed [28] opines that the systems thinking model gives leaders a deeper understanding of the roles or behaviour of the parts that make up a system. Therefore, in dealing with a complex and dynamic social system, systems thinking becomes crucial to synthesise a problem by seeing things in terms of patterns and relationships. Therefore, the evolution of a systems model for this study is an approach to develop a holistic understanding of multi-stakeholders' influence on health and safety risk management in building construction projects.

A system is the concept where one level can be appropriately regarded as nested within another level. The levels are characterised by emergent properties that are irreducible and represent constraints on the degree of freedom of components at the level below. Hierarchies, under the system theory, are characterised by control and communication processes operating at the interfaces between levels [29]. This concept literally creates an environment in which all the system, subsystems and super system are linked together to achieve the overall project goal. Construction projects are manifestations of the hierarchy's concept of a system in terms of the arrangement of subsystems, systems and super systems [30]. A construction project operates in a large/super system such as construction industry in Tanzania. The construction project also comprises different stages or levels in its project lifecycle with highly diverse stakeholders from inception to completion and then to project commission. These levels or stages include briefing, design, procurement, construction and commissioning whereby each system acts as a sub-system. The adopted hierarchy concept of system illuminates how different stages of construction projects are interrelated in terms of stakeholders' participation in risk management.

Additionally, a system can be either closed or open. Closed systems are those that do not interact with their environment. On the other hand, open systems dynamically exchange information with their environment in the form of feedback loop [31]. Construction projects have been regarded as an open system. This open system is affected by and exchanges information with the environment. Moreover, project stakeholders are guided and regulated by different regulatory bodies, professional societies, policies and regulations, political systems economic and market forces in the briefing, designing, procurement and construction processes.

Therefore, the evolution of the system model for this study is an approach to develop a holistic understanding of multi-stakeholder's influence on health and safety risk management in building construction projects. The system model becomes crucial to synthesise a problem by seeing things in terms of patterns and complexity of interrelationships of stakeholders, their roles and responsibilities, communication and how they influence health and safety risk management.

3. Methodology

The main issues were to assess first the factors of the influence of the stakeholder's participation on health and safety risk management and then using case study approach to test those factors in the real projects. The study adopted mixed method whereby both the quantitative and qualitative research approaches were used.

The first part of the study was quantitative method based on questionnaire survey for clients, architects, engineers, quantity surveyors, contractors and site workers. Questionnaires were administered by research assistants. Out of 100 questionnaires distributed, 84 (84%) were fairly filled and returned. Respondents were randomly selected. Ten factors were identified from the literature and respondents were required to rank the way they influence them by using a five-point Likert scale. In a Likert scale, they were asked to respond to each of the factors. The ratings used were: 1—Very poor, 2—Poor, 3—Moderate, 4—Good, and 5—Very good. Inferential statistics were used for the analysis of the data, for the initial stage of the data analysis as indicated in **Table 1**. The hypothesized value is the middle of the used Likert scale which is equal to 2.5. The factors which have mean score above 2.5 were considered to be relevant and were validated in the case study.

The second part of the study was qualitative method whereby case study was adopted. The main purpose of this phase was to develop a clear understanding of how project stakeholders can influence health and safety risk management in construction project based on the factors from survey. Three large building construction projects were selected, and the unit of analysis was stakeholders involved

S/N	Factors	N	Mean score	Rank
1	Role performed by stakeholders	80	4.3	1
2	Individual knowledge and experience	80	4.1	2
3	Stakeholders' power attributes	80	4.1	2
4	Condition of contract	80	4.0	4
5	Nature of health and safety risk	80	4.0	4
6	Communication mode	80	3.9	6
7	Individual perception	80	3.8	7
8	Health and safety regulations	80	3.8	7
9	Professional by-laws	80	3.8	7
10	Procurement regulations	80	3.8	7
	Average mean		3.96	
	Cronbach's alpha: 0.97 (moderate reliability) Average inter-item correlation: 0.53			

Table 1.
The ranking of mean scores (MSs) on the factor influences stakeholders on health and safety risk management.

in these projects. These stakeholders are clients, project managers, architects, engineers, quantity surveyors, site managers, health and safety committee and construction workers. Interview and observations were used for data collection within the case studies. Case studies were selected based on the size of the project (large project, number of employees more than 20 on the sites and falling under design-bid-build procurement method (thus, project with clear separation of project phases, briefing, design, procurement (bidding) and construction). The following projects were chosen:

Project A—The construction of a two-storey warehouse. The scope of the work included demolition of the existing building and construction of the double-storey building which comprised stores and office accommodation. The contract value of the project is Tsh 3,350,200,000 = $ 1,298,527.13 USD as per June 2017 exchange rate with contract period of 18 months. The project was procured using design-bid-build method whereby the client was private sector. The project had a safety department well equipped with safety equipment and first-aid facilities with full-time safety officer. The client had safety policy and was involved in planning of the various safety features in the project.

Project B—The construction of five-storey maternity ward in one of the hospitals. The project had a contract period of 30 month with contract value of Tsh 7,412,470,000 = $4,547,527 USD. The client was a government institution and design-bid-build procurement method was used. The site had safety department and four safety officers with one safety coordinator. The client, consultant team and contractor had safety policies. The client and consultant team selected contractor based on safety merits.

Project C—The construction of a 26-storey building comprising residential apartments, offices and car park accommodation located in Dar es Salaam. The contract value of project was Tsh 132,254,917,029 ≈ $80,889,856 USD as per June 2017 exchange rate with a contract period of 162 weeks. The project was procured using the design-bid-build contract with client being the government institution. The client employ project manager and clerk of the work who stayed at the site full time. One of the responsibilities was to ensure that the contractor has adhered to the health and safety issues. The consultant team and contractor had safety policies.

4. Data presentation and discussion

Table 1 indicates that all factors that influence stakeholders' participation on health and safety risk management have mean score above the midpoint of 2.5, with an average mean of 3.96. The leading factor was the role performed by stakeholders in the project. Thus, indicating that health and safety risk management is embedded in the role performed by stakeholders, individual knowledge and experience and power attributes were ranked second. Other factors include nature of health and safety risk, mode of communication, individual perception, health and safety regulations, professional by-laws and procurement regulations. This implies that all factors are relevant for stakeholder to influence health and safety risk management performance in construction projects. Based on this finding, the factors were tested in the three construction projects through systems thinking approach.

5. Systematic view of stakeholders' influence on health and safety risk management

To understand the influence of stakeholders on health and safety risk management, one should look at the roles performed by each stakeholder in each project stage (briefing, design, procurement and construction) and how their role influences health and safety management. The stakeholders considered are those whose performances play an important role in determining project success. These stakeholders include the project client, project management consultants (project managers, architects, engineers and quantity surveyors) and project contractors, sub-contractors and workers as well as regulators and legislators in the legal system. Stakeholders' participation in health and safety risk management during the different stages is analysed and presented in **Figures 1–4**. Blue shadow indicates the most active stakeholder at the stage considered. The grey shadow indicates the stakeholders whose decisions at the previous stages have consequences during the actual stage.

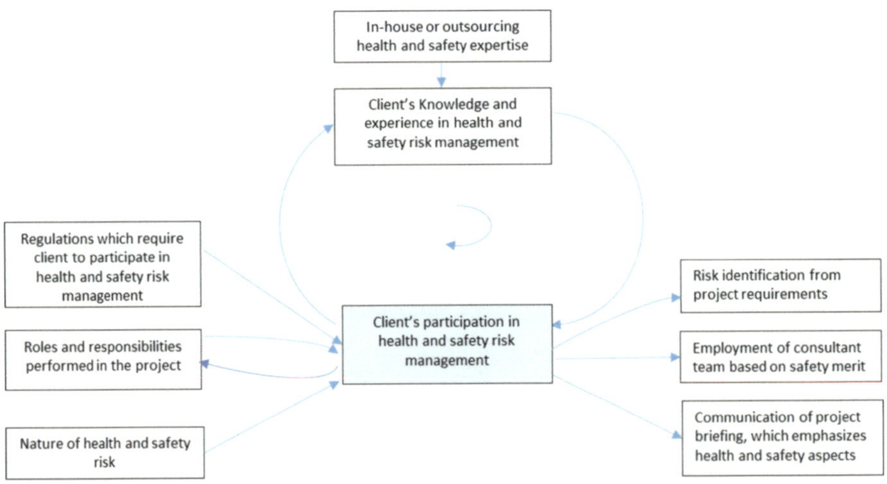

Figure 1.
Stakeholders' participation in health and safety risk management during the briefing stage, according to the cases analysed.

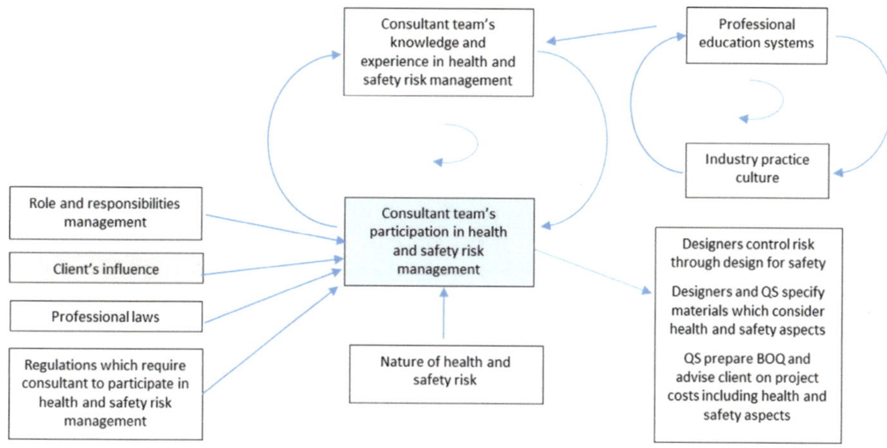

Figure 2.
Stakeholders' influence on health and safety risk management during the design stage, according to the cases analysed.

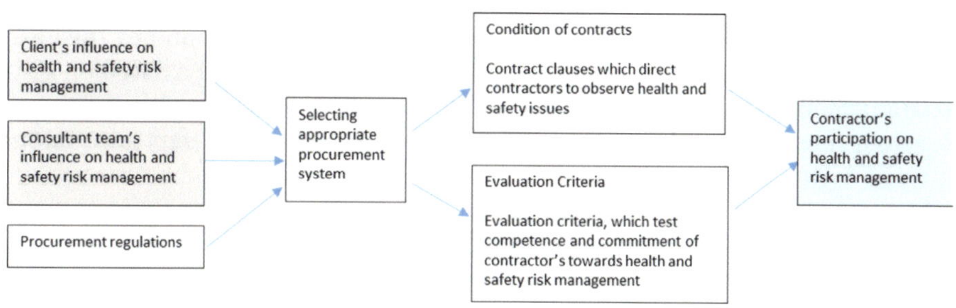

Figure 3.
Stakeholders' influence on health and safety risk management during procurement stage, according to the cases analysed.

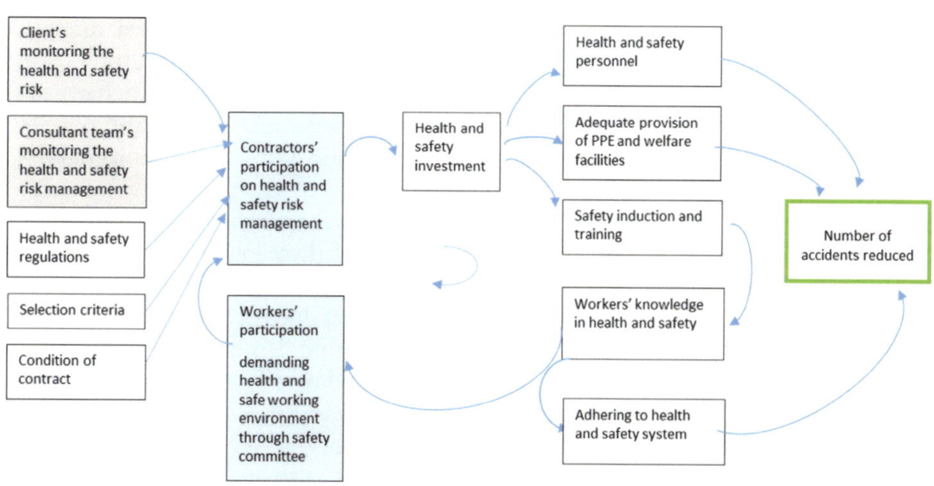

Figure 4.
Stakeholder's influence on health and safety risk management during the construction stage, according to the cases analysed.

5.1 Briefing/inception stage

During the briefing stage of projects, the clients are observed as the main actors in all decisions concerning project initiation and implementation. Thus, clients were directly involved in initiating projects, setting up project requirements, employment of the consultant teams and communicating project requirements to the consultant teams. During the inception stage of the project, the study further reveals that all the clients of three projects were involved in the identification of hazards during project requirement set-up. In the two projects (A and B), they identified hazards such as noise and dust whereas in Project C, the nature of the site was confined and, therefore, hazards issues were also considered at this stage. Moreover, clients employed consultant teams based on their competency; in Project A, the consultants were employed based on previous experience with client and health and safety merit, whereas in Projects B and C, the consultant teams were employed based on Public Procurement Act (PPA) regulations of 2005, primarily because these were public projects. The implication of this orientation is that in Project A, the client was free to make a better choice of the consultant because he/she was free to use any appointment method. This was particularly instrumental because he/she had documented the practical experience of the shortlisted consultants. Of course, one cannot rule out bias or decision made on the basis of vested interest, particularly because in private projects, clients are not bound by the PPA provisions. At the same time, where the consultants were appointed based on general competences, they did not necessarily focus on the aspect of health and safety risk management. Thus, there could be a danger that health and safety issues would be played down. Fortunately, this was not the case in projects B and C.

Furthermore, during the inception stage, the clients had to communicate project requirements to the consultant teams. Multi-channel approaches were used to communicate health and safety information among the consultant teams. Likewise, the clients and the consultant teams collaborated in the identification of hazards and the establishment of control measures in the early stages of the projects. This was observed in all the three projects during the briefing meetings. For example, the client and the consultant team in Project A agreed to change the building outline to bypass the underground electrical cable. In Project C, permission to close one of the free access roads to the site was requested from the authorities. In other words, the project inception stage is very crucial in determining health and safety risk management issues because it is the stage where a client jointly works with consultant teams in a real situation (at the site) where they are actively involved, consult each other, inform and collaborate in identification of hazards, assessment of risks and possible control. Moreover, it is the crucial stage where clients are active in decision-making, therefore quite instrumental in health and safety concern. Moreover, it was noted that clients had experts (in-house or outsourcing) with experience and knowledge in health and safety risk management. The presence of these experts had an impact on client knowledge and consciousness in risk management. This contention underscores the fact that the initiator/client of a project does not necessarily need to have health and safety knowledge before he/she can build. However, he/she needs to have experts (in-house or outsourcing) to provide guidance on risk management in the briefing stage. This observation is also supported by [15].

Figure 1 illustrates that, regulations, role performed by clients and the nature of health and safety risks are guiding client to influence health and safety risk management. This indicates that regulations should assign client responsibilities for health and safety risk management and specify the role the client should play.

It further revealed that to achieve health and safety risk management, knowledge and experience must exist among individuals/groups at a particular point.

While clients participate in health and safety risk management, they acquire more experience and knowledge, hence forming the feedback loop of acquisition of knowledge and experiences.

5.2 Design stage

This study has established that the consultant team, to some extent, has influenced health and safety risk management. It occurs predominantly in the control of risk during design, consideration of health and safety items in the BOQ, and the provision of the budget for health and safety matters. For example, in Project A, the designers were involved in risk control through the design for safety, while it was not the case in projects B and C. In Project A, design for safety was required by the client. In many instances, design decisions can be regarded as the 'source' of health and safety risks in the construction industry; therefore, they ought to be checked and addressed at design stage. However, the design for safety was a major challenge across the projects analysed. For example, in Project A, design for safety was only implemented on the concept outline, rather than being fully integrated in detailed design and material specifications, while in Projects B and C, it was not implemented at all.

Consultant teams in all three projects acknowledged to have limited knowledge on designing for safety. This was also supported by the findings from the review of curricular of higher education of architects and engineers where there were no modules covering health and safety aspects. The other reasons for not considering health and safety during design were associated with conventional opinion among professionals that safety is contractors' responsibility and with lack of regulations supporting designers (architects and engineers) to apply design for safety. Nonetheless, in Project C, there was at least a provision for special professional indemnity insurance that covered accidents that may arise because of faults in the design. The presence of professional indemnity insurance indicates that the designers were committed to legal liabilities for health and safety risks. On the other hand, quantity surveyors participated actively in the preparation of BOQs and cost estimate in all the three projects. It was noted that financial provision for health and safety risk management was made in the preliminary items in BOQ and contingency. At this stage, the consultant teams largely provide advice, leaving key decisions to the clients.

Figure 2 provides summary of the relationship of components that influence health and safety risk management during the design stage.

Figure 2 shows that the opportunities provided by the roles played by the team, client engagement, professionals' by-laws and regulations, the nature of health and safety risk are necessary for consultant the team to influence health and safety risk management during the design stage. It is further revealed that in order to influence health and safety risk management, relevant knowledge and experience are required. Knowledge and experience can be taped from industry practice culture and higher learning institutions. Therefore, it can be noted that the more the consultant team participate in health and safety risk management, the more knowledgeable and experience they will be, hence they will participate more on health and safety risk management. This indicates the feedback loop. On the other hand, furthermore, the more the professional education system includes health and safety knowledge, the more the industry will tap this knowledge through practice.

5.3 Procurement stage

During the procurement stage, the main emphasis of the study was to evaluate the contractors' capabilities and commitment towards risk management. Whereas in Project A, the contractor was employed through the shortlisting method, in Projects B and C, the contractors were employed based on the competitive bidding method in accordance with the criteria stipulated in the Public Procurement Regulations of 2005. In this regard, in Projects B and C, the clients had to adhere to the stipulated procedure and guidelines. Major criteria for selecting consultants and contractors were based on the general competences; therefore, as already noted, knowledge on health and safety was not an issue. As a consequence, it was possible to select a contractor with unproven or unsafe track record or practice within health and safety risk control.

Apart from established regulations, however, it was observed that the additional criteria beyond those established in the regulations were used in Project B. The client and the consultant team set criteria such as site safety management, provision of Personal Protective Equipment (PPE) to the workers and provision of first-aid facilities. The criteria were derived from the site visits of the client and the consultant team and introduced to the bidders for ongoing construction projects. This approach offered a way to test information on the ground to ensure correct decisions were made. It is particularly so, if such visits are made in advance, without knowing the prospective contractors (before the contractors are selected).

The study established the link between the procurement process and contractors' competence and commitment to health and safety risk management. The study has further established that procurement process is influenced by the client, the consultant team and existing regulation as indicated in **Figure 3**.

Figure 3 illustrates in which way the employment of a competent and safety-conscious contractor depends on the client's and the consultant team's participation as well as the procurement regulations. Indeed, if the client's and consultant team's level of participation in risk management increases, the number of pre-qualified criteria involved for selecting an appropriate contractor will also increase. On the other hand, conditions of contracts with specific clauses which address health and safety risk management will also increase the contractor's commitment towards enhancing effective health and safety management practice.

5.4 Construction stage

During the construction stage, contractors had many health and safety responsibilities as they were involved in actual activities at the sites. They employed health and safety personnel, provided PPE and other welfare facilities to the workers. They also had to assess, communicate and control risks on-sites. It was noted that one safety officer was employed in Project A, whereas four and six officers were employed in Projects B and C respectively. Project A had only one safety officer possibly because of the nature of the project in terms of size, location, site configuration and number of employees. The Project A site had a total of 40 employees and the project comprised two floors. In Projects B and C, projects were more complex with 5 and 26 floors and employed approximately 400 and 500 workers respectively. What is critical here, however, is that the presence of safety officers is an indication that the contractors were committed to risk management. Safety officers play a critical role; they work on behalf of site managers to identify risk, communicate risk to workers and control risk.

During the construction stage, the contractors provided safety induction training to workers in all the three projects. In Project A, the client collaborated with the contractor to provide safety induction training, whereas in Projects B and C, clients required the contractors to provide health and safety training to the workers and submit training reports to them. The conclusion is that clients can play specific role in ensuring that workers are well informed about health and safety risks and overall risk management efforts at building sites. This finding is consistent with that of study done by [32] that multiplicity of communication channels, methods and different stakeholders play out simultaneously in the communication of risk information within construction sites.

In all the three projects, the clients and consultant teams participated in regular inspections during the construction stage. Most importantly, the inspections were done randomly, without prior notification; issues observed during site inspection included the contractors' compliance in health and safety matters on-sites. During inspection, in all three projects, compliance certificates were issued and shortfall notices or fines were imposed on those who did not comply. Accordingly, the consultant team issued warning letters to Project B where defaults were noted. Sometimes, photographs were taken as evidence and site inspection reports were presented in the site monthly meetings. Such steps helped to boost safety risk management.

Furthermore, it was observed that in the three projects, all stakeholders were closely linked together through regular site meetings which were held monthly. Health and safety on-site were among the main agenda of these meetings. This implies that these meetings are monitoring tools for health and safety risk management performance where some of the stakeholders were informed, consulted, advised and involved in decision-making. This finding is consistent with that of study [33] that different stakeholders have different sources of power such as technical expertise; legitimate, political position; resources information; reward and coerce power which influence risk management in construction sites.

It was also noted that Projects B and C had safety committee meetings. Safety committee meetings involved the workers' representatives, the contractors' safety team and the client' representatives. The workers' representatives were elected by workers themselves from each trade. The elections of safety representatives exemplified political power among the workers. It was observed that these meetings were partly interactive, especially when the workers' representatives raise their concern towards health and safety issues on-site.

One can say that contractors still bear relatively high responsibilities of health and safety risk management. However, contractor fulfilment of these responsibilities depends on: how he/she was procured, the consultant management and supervision, client's demands and the existing regulations as indicated in **Figure 4**.

Figure 4 indicates that contractor's participation in health and safety risk management in construction projects requires client's monitoring system, consultant's communication and monitoring system, health and safety regulations which assign specific responsibilities to contactors, conditions of contract which direct the contractor to observe health and safety aspects and the evaluation criteria which test the competencies in health and safety issues.

It also further illustrates that, there is a causal relationship between contractor's participation in health and safety risk management and workers' participation. Thus, if the contractor is committed to health and safety risk management, he/she can employ health and safety personnel, provide proper PPE to the workers and provide safety induction to the workers. The presence of safety personnel such as safety managers and safety officers on-site is essential for communicating safety information to the workers. Therefore, if the number of safety personnel increases, the amount of safety induction and refreshers training would also increase. These would, in turn, boost the workers' knowledge and, hence, they would adhere to safety practices such as wearing PPE all the time and proper housekeeping. Workers' adhering to safety risk management would eventually reduce the number of accidents on construction sites. On the other hand, the workers' knowledge would influence them to demand for better working environment through safety committee meetings which would, in turn, influence the contractor's investments in risk management.

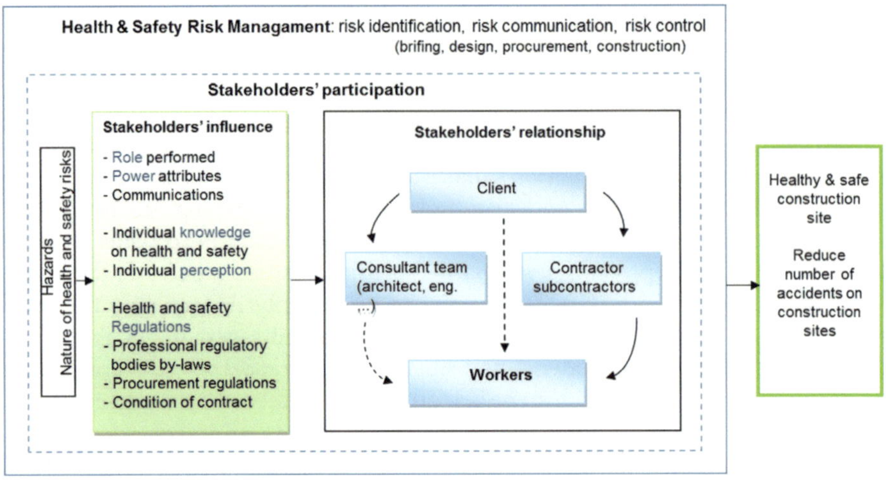

Figure 5.
Framework for stakeholders' participation in health and safety risk management in construction projects in Tanzania.

5.5 Development of final framework

The final framework was developed by considering the findings from survey and validated in multiple case studies with the aim of enhancing clarity and effectiveness. The framework consists of the following features: factors influencing stakeholders' participation and stakeholders' relationships in the health and safety risk management (hazard-risk identification, risk communication and risk control). The output of the framework is to reduce the number of accidents on construction sites.

Figure 5 indicates that when stakeholders participate in health and safety risk management, they are influenced by their roles and responsibilities in the project, the communication process in place, the stakeholders' power attributes and individual knowledge on health and safety issues as well as by health and safety regulations and professional bodies' by-laws. Therefore, to ensure a smooth risk management process and effective stakeholders' influence on risk management, the balance of those factors must be considered. **Figure 5** further indicates that when stakeholders participate, they create a link chain relationship. This 'link chain' participation relationship is in conformity with the concept of interconnectedness of system thinking. Thus, each stakeholder's involvement in a system has a critical role in influencing and making the system work. Non-performing of one stakeholder can break the chain, hence making the system fails to work. The goal of risk management is to reduce the number of accidents in construction sites. A mechanism to ensure each stakeholder's influence one another is paramount.

6. Conclusions

This research has investigated multi-stakeholders' influence on health and safety risk management in three large construction projects in Tanzania. There is an empirical evidence showing that different project's stakeholders such as clients, architects, engineers, quantity surveyors, contractors and workers are able to participate in health and safety risk management at different phases/stages of the project and in different ways. However, the holistic view over their relationships and actions is needed to make the risk control process effective. This study has demonstrated a 'link chain' relationship reflecting the stakeholders' influence on health and safety risk management. While this chain relationship is evident, still there can be some gaps or vulnerabilities making the system not work effectively. Therefore, this study proposes the framework showing important factors and links to improve stakeholders' influence on health and safety risk management.

The graphs given in **Figures 1–5** concerning the following stages of the construction projects show the changing responsibilities of the main stakeholders concerning health and safety and the importance to secure the continuity of the health and safety risk management process through all the stages because the risks on construction sites depend on the decisions made earlier. It is noted that the role and responsibilities performed by stakeholders give potential opportunity for them to influence health and safety risk management. Thus, the health and safety risk are embedded in their role and responsibilities performed in the project. A clear understanding of the role and responsibilities of each stakeholder in the project is important. Also, knowledge and skills of health and safety risk management are very important and a pre-requisite for clients, designers, QS, contractors and workers' participation process. This knowledge and experience are obtained from the construction industry practice and training institutions. Therefore, health and safety modules should be emphasised in curricula of training institutions that produce professionals such as architects, engineers and quantity surveyors. The existing health and safety regulations and by-laws from professional registration bodies, procurement regulations and condition of contracts play an important role to influence stakeholders to participate in health and safety risk management. Thus, if the regulations are strong and supported by strong enforcement, the industry practice culture will also change in a positive way. The existing health and safety policies and regulations need to be reviewed to influence stakeholders to participate in health and safety risk management much more effectively and substantively.

The pattern of relationships between different stakeholders and the capacity of certain control actions/tools were analysed using the results of case studies for all three projects. For the individual project, the system dynamics approach could be applied on the more detailed level tracing the specific pattern relevant for the actual stakeholders' participation successes and failures to build up effective and robust system for healthy and safe construction site.

Author details

Sarah Phoya[1]* and Krystyna Pietrzyk[2]

1 Ardhi University, Dar es Salaam, Tanzania

2 Chalmers University of Technology, Gothenburg, Sweden

*Address all correspondence to: sarahphoya@gmail.com

References

[1] Lingard H, Rowlinson SM. Occupational Health and Safety in Construction Project Management. London: Taylor & Francis; 2005

[2] Gambatese JA, Behm M, Rajendran S. Designs role in construction accident causality and prevention: Perspectives from an expert panel. Journal of Safety Science. 2008;**46**:675-691. DOI: doi.org/10.1016/jssci 2007.06.010

[3] Hallowell R, Hinze JW, Baud KC, Wehle A. Proactive construction safety control: Measuring, monitoring, and responding to safety leading indicators. Journal of Construction Engineering and Management. 2013;**139**:10. DOI: 10.1061/(ASCE)CO.1943-7862.0000730

[4] Palencher ML, Heath RL, Hocke T. Corporate (social) responsibility, risk management and communication. In: Ihlen O, Bartlett J, May S, editors. Handbook of Communication and Corporate Social Responsibility. West Sussex: Wiley Blackwell; 2011

[5] Charles M, Pillay J, Ryan R. Guide to Best Practice for Safer Construction: Literature Review 'From Concept to Completion. Brisbane, Cooperative Research Centre for Construction Innovation, for Icon. Net Pty Ltd; 2007

[6] Zhang S, Sulankivi K, Kiviniemi M, Romo I, Eastman CM, Teizer J. BIM-based fall hazard identification and prevention in construction safety planning. Journal of Safety Science. 2015;**72**:31-45. DOI: 10.1016/j.ssci.2014.08.001

[7] International Labour Office. Safety and Health in Construction Code of Practice. Geneva: ILO; 1992

[8] Behm M. Linking construction fatalities to the design for construction safety concept. Journal of Safety Science. 2005;**43**(8):589-611. DOI: 10.1016/j.ssci.2005.04.002

[9] Atkinson AR, Westall R. The relationship between integrated design and construction and safety on construction projects. Construction Management and Economics. 2010;**28**(9):1007-1017. DOI: 10.1080/01446193.2010.504214

[10] Sacks R, Whyte J, Swissa D, Raviv G, Zhou W, Aviad Shapira A. Safety by design: Dialogues between designers and builders using virtual reality. Construction Management and Economics. 2015;**33**(1):55-72. DOI: 10.1080/01446193.2015.1029504

[11] Chileshe N, Dzisi E. Benefits and barriers of construction health and safety management (HSM) perceptions of practitioners within design organisations. Journal of Engineering, Design and Technology. 2012;**10**(2):276-298. DOI: 10.1108/17260531211241220

[12] Smallwood JJ. The role and influence of clients and designers in construction health and safety. In: First European Conference on Construction Health and Safety Coordination in the Construction Industry. 2008. pp. 21-22

[13] Bikitsha L, Haupt T. Impact of prefabrication on construction site health and safety: Perceptions of designers and contractors. In: The ASOCSA 6th Built Environment Conference. Sandton, South Africa; 2011. pp. 196-214

[14] Lingard H, Blismas N, Cook T, Cooper A. The model client framework, resources to help Austrian government agencies to promote safe construction. International Journal of Managing Projects in Business. 2009;**2**(1):131-140. DOI: 10.1108/17538370910930554

[15] Nnedinma U. An appraisal of the barriers to client involvement in health and safety in Nigeria's construction industry. Journal of Engineering, Design and Technology.

2017;**15**(4):471-487. DOI: doi.org/10.1108/JEDT-06-2016-0034

[16] Smallwood J. The impact of a client contractor health and safety (H&S) programme on contractor H&S performance. Procedia Engineering. 2017;**196**:996-1002. DOI: 10.1016/j.proeng.2017.08.041

[17] Wells J, Hawkins J. Briefing: Promoting construction health and safety through procurement. Proceedings of the Institution of Civil Engineers—Management, Procurement and Law. 2011;**164**(4):165-168. DOI: 10.1680/mpal.10.00001

[18] Freeman E. Strategic Management: A Stakeholder Approach. Boston, MA: Pitman Publishing; 1984

[19] Mitchell RK, Agle BR, Wood DJ. Toward a theory of stakeholder identification and salience: Defining the principle of who and what really counts. Academy of Management Review. 1997;**22**(4):853-887. DOI: 10.5465/amr.1997.9711022105

[20] Clarkson M. A stakeholder framework for analyzing and evaluating corporate social performance. Journal of Academy of Management Review. 1995;**20**(1):92-117. DOI: 10.5465/amr.1995.9503271994

[21] Leung M, Olomolaiye P. Risk and construction stakeholder management. In: Chinyio EA, Olomolaiye P, editors. Construction Stakeholder Management. Chichester: Wiley-Blackwell; 2010

[22] Olander S, Landin A. A comparative study of factors affecting the external stakeholder management process. Journal of Construction Management and Economics. 2008;**26**(6):553-561. DOI: 10.1080/01446190701821810

[23] Ward S, Chapman C. Transforming project risk management into project uncertainty management. International Journal of Project Management. 2003;**21**:97-105

[24] Smith N, Merna T, Jobling P. Managing Risk in Construction Projects. London: Blackwell; 2006

[25] Klemetti A. Risk management in construction networks. 2006. Available from: www.lib.tkk.fi/reports2006 [Accessed: Feb 29, 2016]

[26] Senge PM. The Fifth Discipline: The Art and Practice of Learning Organisation. London: Random House Business Book; 2006

[27] Checkland P. Systems Thinking, Systems Practice. Chichester: John Wiley & Sons; 1990

[28] Reed GE. Systems thinking and senior leadership. In: Grandstaff RM, Sorenson G, editors. Strategic Leadership: The General's Art. USA: Management Concepts; 2009. pp. 77-93

[29] Leverson G. Engineering a Safer World: A Systems Thinking Applied to Safety. Massachusetts: MIT Press; 2011

[30] Walker A. Project Management in Construction. London: Wiley; 2007

[31] Flood RL, Jackson MC. Creative Problem Solving: Total Systems Intervention. New York: John Wiley & Sons; 1991

[32] Phoya S, Eliufoo H. Perception and the uses of stakeholders' power on health and safety risk management in different stages of construction projects in Tanzania. Journal of Civil Engineering and Architecture. 2016;**10**:955-963. DOI: 10.17265/1934-7359/2016.08.012

[33] Phoya S. The practice of communication of health and safety risk information at construction sites in Tanzania. International Journal of Engineering Trends and Technology. 2017;**47**(7):385-393. DOI: 10.14445/22315381/IJETT-V47P26

Chapter 5

A Guide for Risk Management in Construction Projects: Present Knowledge and Future Directions

Pinar Irlayici Cakmak and Ecem Tezel

Abstract

Construction projects are well known to be prone to a high level of risk that cannot be ignored but can be managed. Researchers have studied numerous aspects of risk management including identification, analysis/assessment, response and control. Despite the fact that studies focused on risk management in construction projects have been increasing, there seems to be a limited number of published studies that summarize what has already been presented in the literature. In this regard, this chapter aims to present the existing literature on risk management from a holistic perspective and provide a guide for future directions. With this aim, a systematic literature review has been undertaken by presenting the areas focused on by researchers as well as neglected ones, by indicating the trends in research through the years and by discussing research gaps for potential studies.

Keywords: construction projects, future directions, research trends, risk management, systematic review

1. Introduction

Construction projects involve participants from different specialties working together which makes the cooperation among them designed around extensive, disparate and interrelated processes [1]. Such complexity is also increased by other external factors such as political, legal, cultural, technological and financial, which resulted in project risk. Project Management Institute (PMI) defines project risk as "an uncertain event or condition that, if it occurs, has a positive or negative effect on a project objective. ... there will be a consequence on the project cost, schedule, or quality" [2]. Due to the increasing size and complexity, a wide variety of risks impact the successful completion of the construction projects. In other words, risks are threats to project success [3]. Despite trying to eliminate all the risks in construction projects is impossible, a formal risk management process is required to manage them effectively [4].

In this regard, a systematic risk management process can help construction companies to identify not only the involved risks of projects but also to mitigate impacts of those uncertainties in different phases of projects [5]. The term "risk management" can be broadly defined as work that classifies, analyses and responds to unpredictable risks that exist in the processes of project implementation [6]. Risk management is about defining sources of uncertainty (risk identification), estimating the consequences of uncertain events/conditions (risk analysis), generating response strategies in the light of expected outcomes and, finally, based on the feedback received on actual outcomes and risks emerged, carrying out identification, analysis and response generation steps repetitively throughout the life cycle of a project to ensure that the project objectives are met [7]. Briefly, a traditional risk management process consists of risk identification, risk analysis or assessment, risk response or mitigation and risk monitoring and control [2, 8].

The initial step of risk management is risk identification. Risk identification is the process of identifying individual project risks as well as sources of overall project risk and documenting their characteristics [2]. Although it is difficult to define and measure, it is very important to identify potential risks as early as possible. In order to manage risks properly, risk identification should be performed along with the project's initiation stage. Construction companies usually benefit from risk checklists [9, 10] and risk breakdown structures [9, 11, 12] for the identification.

Risk analysis/assessment is the process that focuses on evaluating and seeking the likelihood in which potential risks in the risk identification stage may occur [13] and it is implemented by two approaches: qualitative risk analysis and quantitative risk analysis. In qualitative risk analysis process, the main focuses are rating and prioritizing individual project risks for further analysis or action by assessing their probability of occurrence and severity of consequence/impact as well as other characteristics [2, 14]. On the other hand, quantitative risk analysis process focuses on numerically analyzing the combined effect of identified individual project risks and other sources of uncertainty on overall project objectives [2]. Researchers employed Delphi [15–17], AHP/fuzzy AHP [10, 17–23] and Monte Carlo simulation [24–26] to assess risks in their studies.

Risk response process consists of developing options, selecting strategies and agreeing on actions to address overall project risk exposure, as well as to treat individual project risks, and finally implementing agreed-upon risk response plans [2]. Dealing with negative consequences, risk response is also referred to as risk mitigation, risk elimination, risk prevention and risk reduction [8]. Appropriate risk response strategies must be selected to reduce risk exposure once the risks have been identified and analyzed [27]. Researchers widely agree that the selection of risk response strategy is an important issue in project risk management [28–30]. These strategies are avoiding, reducing or accepting project risks.

Risk monitoring and control process is the process of monitoring the implementation of agreed-upon risk response plans, tracking identified risks, identifying and analyzing new risks and evaluating risk process effectiveness throughout the project [2]. This step ensures that all information generated by risk management process is captured, used and maintained throughout the construction period [31].

The subject of risk management in construction projects has been increasingly studying since the 1980s. Most of these studies have focused on how risks are identified or analyzed/assessed in different countries such as Australia [32, 33], China [23, 34, 35], Ghana [36], Hong Kong [37, 38], India [39, 40], Indonesia [41, 42], Italy [43], Korea [44], Malaysia [31, 45], Mexico [46], New Zealand [47, 48], Nigeria [49, 50], Poland [51], Singapore [52, 53], Spain [54], Sri Lanka [55], Tanzania [56], the United Kingdom [57, 58], the United States of America [59, 60], Vietnam [61, 62] and Zambia [63]. These studies mostly used survey/interviews or case studies. Additionally, researchers proposed that various theoretical and mathematical models are also proposed for managing risks effectively and efficiently.

While literature is rich in papers addressing risk management in construction projects, few papers have researched what has already been presented. Edwards and Bowen's [64] research is one of the exceptional studies which analytically reviews the construction risk literature over the period from 1960 to 1997. Given that two decades have passed since then, it is appropriate to review the progress in risk management research in construction. In this regard, this paper aims to analyze current literature and provide a guide for future studies on risk management in construction projects.

2. Research methodology

To review the risk management literature comprehensively, a twofold procedure was adopted in this study. At first, a systematic literature review was conducted to identify the key scientific contributions in the risk management domain. The findings of the review, then, were statistically synthesized through a meta-analytical approach which is an associated procedure of systematic literature review.

Systematic literature review adopts a replicable, scientific and transparent process that aims to minimize bias through exhaustive literature searches of published studies [65]. On the other hand, meta-analysis helps to analyse these studies by interrelating focused areas and identifying emerging or neglected themes [66].

In this regard, this study has been organized in two stages represented in **Figure 1**.

2.1 Stage 1: systematic literature review

The first stage concentrates on searching for relevant papers using scientific databases, namely, American Society of Civil Engineers (ASCE), Elsevier, Emerald and Taylor & Francis. From these databases, relevant papers were searched in the following construction and built environment-related journals: Automation in Construction (AC), International Journal of Project Management (IJPM), Journal

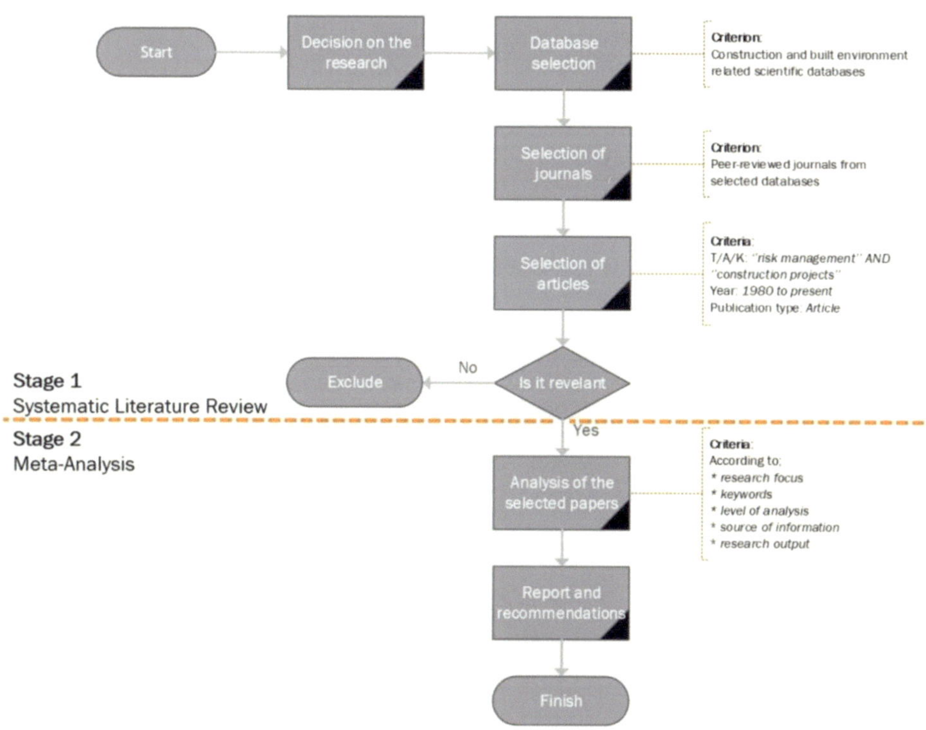

Figure 1.
Stages of the systematic literature review.

of Cleaner Production (JCP), Architectural Engineering and Design Management (AEDM), Construction Management and Economics (CME), International Journal of Construction Education and Research (JCER), International Journal of Construction Management (IJCM), Journal of Civil Engineering and Management (CEM), Journal of Construction Engineering and Management (JCEM), Journal of Management in Engineering (JME), Journal of Legal Affairs and Dispute Resolution in Engineering and Construction (LADR), Journal of Professional Issues in Engineering Education and Practice (PEEP), Journal of Architectural Engineering (JAE), Engineering Construction and Architectural Management (ECAM), Construction Innovation (CI), Journal of Financial Management of Property (JFMP), Facilities (F), Built Environment Project and Asset Management (BEPAM), Journal of Facilities Management (JFM), International Journal of Building Pathology and Adaptation (JBPA) and Management Decision (MD).

The keywords for searching were designated as "risk management" and "construction projects," and these keywords were searched in title/abstract/keyword fields of the selected journals in the time period between 1980 and 2018. At this point, a total of 471 papers, excluding book reviews, forums and editorials, were retrieved for further analysis. Eventually, 247 papers were considered as the most relevant to the research aim and were subject to a detailed review.

2.2 Stage 2: meta-analysis

In the second stage, a meta-classification framework, adopted from Betts and Lansley [66], was designed as presented in **Table 1**. Accordingly, the framework has nine categories, such as year, scientific database, journal, keyword, research focus, level of analysis, source of information, research output and future directions with their related subcategories.

Category	Subcategory
Year	Publication date of the article
Scientific database	ASCE, Elsevier, Emerald, Taylor & Francis
Journal	Name of the journals
Keyword	"Risk management" and "construction projects"
Research focus	Risk identification Risk assessment/analysis Risk evaluation Risk response Risk monitoring and control
Level of analysis	Project level Firm level Sector level
Source of information	Review Case study Survey/interview
Research output	General insights and descriptions Statistical results Theoretical model Mathematical model Experimental/prototype model
Future directions	Future research identified in the articles

Table 1.
Research framework.

The 247 papers were analyzed according to this framework and classified by one of these subcategories. In some cases, a paper may be classified in multiple subcategories, resulting in the sum of the papers distributed among the subcategories exceeding the number of papers analyzed.

3. Data analysis and results

Risk management in construction projects was analysed according to the meta-classification framework given in **Table 1**. It is found that 247 papers have been published on "risk management" in the specified time period in the widely accepted construction and built environment-related peer-reviewed journals.

Table 2 shows the chronological distribution of the selected papers by a 5-year time period. Accordingly, risk management subject shows an increasing tendency over the years. In addition, half of these papers have been published in the ASCE's Journal of Construction Engineering and Management.

Table 3 presents the research focus of the published papers over the years. As given in **Table 3**, research focus was classified into ten categories. These categories include four processes of risk management and their multiple combinations. It is noticeable that researchers studied the risk management subject whether discussing

Database	Journal	≤1995	1996–2000	2001–2005	2006–2010	2011–2015	>2015
Elsevier	AC				3	4	
	IJPM	1	1	2	4	4	1
	JCP						3
Taylor & Francis	AEDM				1	1	1
	CME	1	4	2	6	2	
	JCER				1	1	
	IJCM			1	1	4	5
	CEM				1	6	3
ASCE	JCEM	6	9	12	27	29	23
	JME			1	2	11	11
	LADR						2
	PEEP		2		3	2	
	JAE		1				
Emerald	ECAM		3	3	2	6	6
	CI						2
	JFMP				4		1
	F					1	2
	BEPAM					2	1
	JFM			1	1	4	1
	JBPA			1			
	MD			1			
Total		8	20	24	56	77	62

Table 2.
Distribution of the selected papers within the time span.

	≤1995	1996–2000	2001–2005	2006–2010	2011–2015	>2015	Total number of papers within time span
RI	1	2	3	8	10	14	38
RA	3	7	11	21	28	23	93
RR		2		4	6	4	16
RMC					3	1	4
RI+RA		3	4	6	12	9	34
RI+RR	1	1	1	3	2	1	9
RA+RR			1	1	2	3	7
RA+RMC						1	1
RI+RA+RR	1	2		5	2	3	13
RI+RA+RR+RMC	1	4	4	8	12	3	32

RI: risk identification, RA: risk analysis/assessment, RR: risk response, RMC: risk monitoring & control.

Table 3.
Analysis of selected papers according to the research focus.

one of the processes, such as risk identification, risk analysis/assessment, risk response and risk monitoring and control, or examining them through a holistic approach. Despite most of the papers focused only on risk analysis/assessment, a considerable amount of papers studied other risk management processes together with risk analysis/assessment subject. Besides, risk response and risk monitoring and control seem to be neglected processes of risk management. Recently, it is seen that these processes have started to be mentioned in risk management-related researches. Still, they do not have similar impact in the construction risk management literature compared with risk identification and risk analysis/assessment processes.

Most commonly used keywords in the analyzed papers are given in **Table 4**. It is not surprising that "risk management" keyword has the largest rate with 28.9%. The second highly rated keyword is risk (financial, political, design, economic, social, legal, safety) with the rate of 23.8%. This is followed by other keywords such as construction management/project management (11.6%), risk assessment including risk prioritization, risk score and risk rating (11.2%); risk analysis (6.0%); risk identification including checklist, risk mapping and risk breakdown structure (5.8%); cost-related issues (4.7%); risk allocation/distribution (2.0%); risk modeling (1.3%); risk response (1.1%); risk control (0.6%); risk mitigation (0.6%); risk perception/attitude (0.6%); risk strategy (0.4%); risk interruptions (0.2%); risk paths (0.2%); and risk propagation (0.2%).

The papers are analyzed according to the study levels as project level, firm level and sector level. **Figure 2** shows the distribution of these levels within the time span. As seen in **Figure 2**, the majority of the papers are studied in the project level. This is resulted from researchers mostly focused how risk is managed within a construction project rather than concentrating on the risks and their effect within a construction company or in the construction sector. Especially beginning with 2006, a huge focus has given to construction risk management studies at the project level. However, there are few studies which concentrate risk management related issues by discussing them through the firm and sector level.

Different sources of information are used in the analyzed papers which were classified as case studies, survey/interviews and reviews. As illustrated in **Figure 3**, among these, case studies and survey/interviews are the leading sources. After 2005, case studies and survey/interviews show a rapid increase. This reveals that secondary data and data collected from sector professionals are the main sources of information

Keywords	Number of papers	Frequency (%)
Risk management	129	28.99
Risk	106	23.82
Construction management/project management	52	11.68
Risk assessment	50	11.23
Risk analysis	27	6.06
Risk identification	26	5.84
Cost related issues	21	4.71
Risk allocation/distribution	9	2.02
Risk modeling	6	1.34
Risk response	5	1.12
Risk control	3	0.67
Risk mitigation	3	0.67
Risk perception/attitude	3	0.67
Risk strategy	2	0.44
Risk interruptions	1	0.22
Risk paths	1	0.22
Risk propagation	1	0.22

Table 4.
Analysis of selected papers according to the keywords.

in the analyzed papers. On the other hand, reviews are relatively less preferred information source for risk management researches.

The main outputs of the papers are shown in **Figure 4**, which were classified into five categories as general insights and descriptions, statistical results, theoretical model, mathematical model and experimental/prototype model. The main contribution is statistical results followed by mathematical model. Since most of the papers adopted a research methodology based on case studies and survey/interviews, it is reasonable that the research output shows a high tendency in statistical

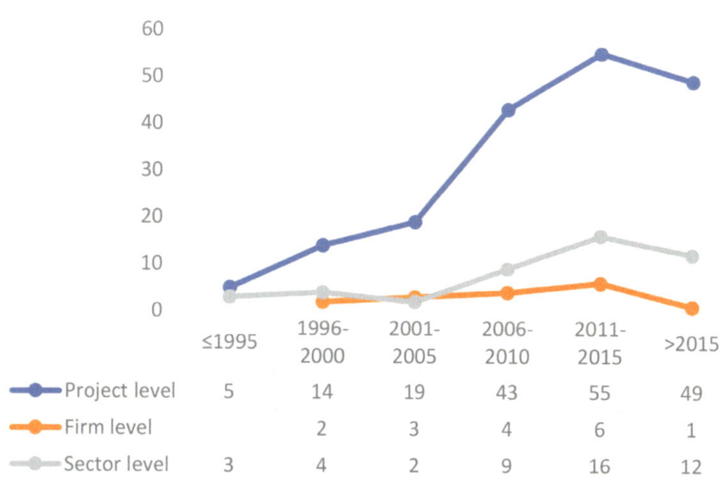

Figure 2.
Analysis of selected papers according to the level of analysis.

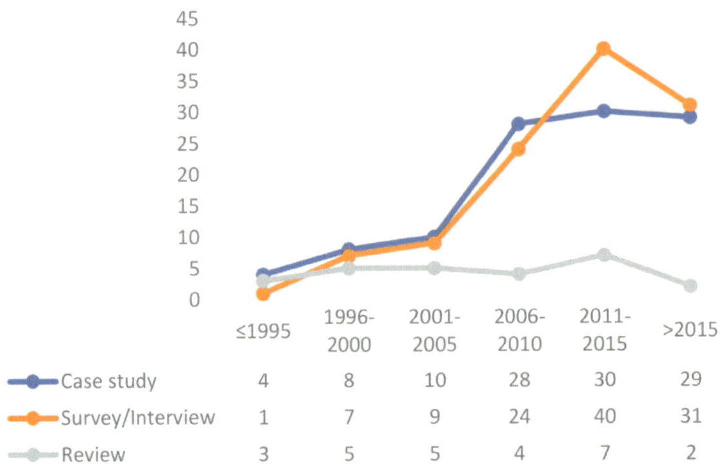

Figure 3.
Analysis of selected papers according to the sources of information.

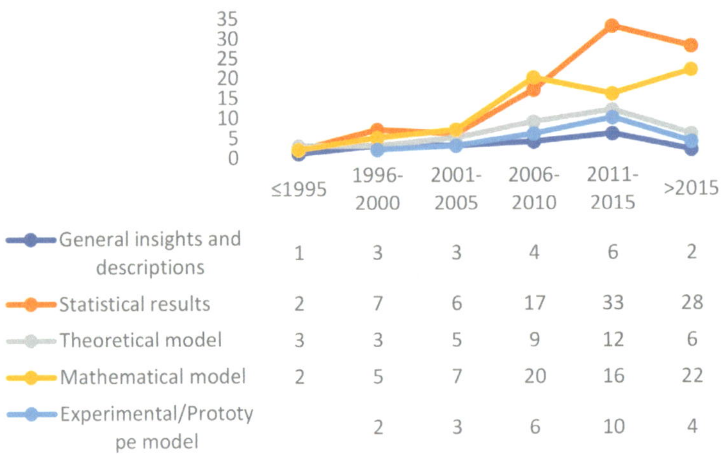

Figure 4.
Analysis of selected papers according to the research output.

results. General insights and descriptions, theoretical models and experimental/prototype models are less adopted methodologies compared with other ones.

4. Conclusion

No construction project is risk-free: risk can be managed, minimized, shared, transferred or accepted; but it cannot be ignored [67]. Construction companies should adopt an appropriate risk management approach not only to complete their projects in compliance with their project objectives but also to keep their competitiveness in the construction industry. Although researchers have drawn huge attention on every step of the risk management process, in this chapter, it is aimed to present the state-of-the-art literature by analyzing research contributions in the risk management domain.

Despite risk management subject found in the literature has reached saturation point, construction researchers have still been studying different aspects of risk management through implementing various research methodologies. A majority of these researches concentrated on one of the risk management processes that is found in risk identification and risk analysis/assessment. On the other hand, the remaining processes of risk management, namely, risk response and risk monitoring and control, are seemed to be neglected.

In the review, highly mentioned risk-related keywords are revealed as risk management; various risk types; construction management/project management; risk assessment including prioritization, risk score and risk rating; risk analysis; risk identification and its methods such as checklist, risk mapping and risk breakdown structure; cost-related issues; risk allocation/distribution; risk modeling; risk response; risk control; and risk mitigation, respectively. As the results showed that risk response and risk monitoring and control are disregarded areas, their related keywords are less mentioned than the other ones.

Since researchers focused how risk is managed within a construction project instead of concentrating on the risks and their effect within a construction company or in the construction sector, a huge number of papers deal with risk management at the project level. Accordingly, studies on risk management at the firm level and sector level seem to be neglected. Besides, as much of the risk management researches in the past decades focused on identification and analysis/assessment of risks within a particular construction project, they mostly adopted survey/interview and case study approaches. This case has resulted in frequent appearance of statistical results as the main research outputs.

To conclude, the review has confirmed that the researchers are directed only on the first two steps of risk management process. In addition to these directions, future studies should also discuss risk response and risk monitoring and control which are the remaining ones. Besides, it is revealed that the literature lacks a comprehensive risk management process. Future studies should adopt a holistic perspective which addresses the risk management process by identifying, analyzing/assessing, responding and monitoring and control from initiation to the completion of construction projects. Similarly, future studies should be directed to risk management-related issues by discussing them at the firm and sector level as well. This systematic review is expected to contribute to the construction profession by enlightening the research gaps in the literature and by providing future directions for potential studies.

Conflict of interest

The authors of this book chapter declare no "conflict of interest."

Author details

Pinar Irlayici Cakmak* and Ecem Tezel
Department of Architecture, Istanbul Technical University, Istanbul, Turkey

*Address all correspondence to: irlayici@itu.edu.tr

References

[1] Burtonshaw-Gunn SA. Risk and Financial Management in Construction. Vol. 2009. Abingdon: Routledge; 2009. ISBN: 978-1-138-24604-1

[2] Project Management Institute. A Guidc to the Project Management Body of Knowledge. 5th ed. Pennsylvania: Project Management Institute; 2013. ISBN: 978-1-935589-67-9

[3] Barber RB. Understanding internally generated risks in projects. International Journal of Project Management. 2005;**23**:584-590. DOI: 10.1016/j.ijproman.2005.05.006

[4] El-Sayegh SM. Risk assessment and allocation in the UAE construction industry. International Journal of Project Management. 2008;**26**:431-483. DOI: 10.1016/j.ijproman.2007.07.004

[5] Chapman C, Ward S. Project Risk Management: Process, Techniques, and Insights. 2nd ed. Chichester: Wiley; 2008. ISBN: 978-0-470-85355-9

[6] Kang LS, Kim S-K, Moon HS, Kim HS. Development of a 4D object-based system for visualizing the risk information of construction projects. Automation in Construction. 2013;**31**:186-203. DOI: 10.1016/j.autcon.2012.11.038

[7] Dikmen I, Birgonul MT, Anac C, Tah JHM, Aouad G. Learning from risks: A tool for post-project risk assessment. Automation in Construction. 2008;**18**:42-50. DOI: 10.1016/j.autcon.2008.04.008

[8] International Organization for Standardization. Risk Management-Principles and Guidelines. Geneva: International Organizations for Standardization; 2009. ISBN: 978-0-580-67571-3

[9] Yildiz AE, Dikmen I, Birgonul MT, Ercoskun K, Alten S. A knowledge-based risk mapping tool for cost estimation of international construction projects. Automation in Construction. 2014;**43**:144-155. DOI: 10.1016/j.autcon.2014.03.010

[10] Zou PXW, Li J. Risk identification and assessment in subway projects: Case study of Nanjing Subway Line 2. Construction Management and Economics. 2010;**28**:1219-1238. DOI: 10.1080/01446193.2010.519781

[11] Zou Y, Kiviniemi A, Jones SW. Developing a tailored RBS linking to BIM for risk management of bridge projects. Engineering, Construction and Architectural Management. 2016;**23**:727-750. DOI: 10.1108/ECAM-01-2016-0009

[12] Hillson D. Using a risk breakdown structure in project management. Journal of Facilities Management. 2003;**2**:85-97. DOI: 10.1108/147259604

[13] Cooper DF, Grey S, Raymond G, Walker P. Project Risk Management Guidelines: Managing Risk in Large Projects and Complex Procurements. Chichester: John Wiley & Sons; 2005. ISBN: 978-0-470-02282-5

[14] Alarcon LF, Ashley DB, de Hanily AS, Molenaar KR, Ungo R. Risk planning and management for the Panama Canal expansion program. Journal of Construction Engineering and Management. 2011;**137**:762-771. DOI: 10.1061/(ASCE)CO.1943-7862.0000317

[15] del Cano A, de la Cruz MP. Integrated methodology for project risk management. Journal of Construction Engineering and Management. 2002;**128**:473-485. DOI: 10.1061/(ASCE)0733-9364(2002)128:6(473)

[16] Xu Y, Yeung JFY, Chan APC, Chan DWM, Wang SQ, Ke Y. Developing a

risk assessment model for PPP projects in China—A fuzzy synthetic evaluation approach. Automation in Construction. 2010;**19**:929-943. DOI: 10.1016/j.autcon.2010.06.006

[17] Shahata K, Zayed T. Integrated risk-assessment framework for municipal infrastructure. Journal of Construction Engineering and Management. 2016;**142**:040150052. DOI: 10.1061/(ASCE)CO.1943-7862.0001028

[18] Youssef A, Osman H, Georgy M, Yehia N. Semantic risk assessment for ad hoc and amended standard forms of construction contracts. Journal of Legal Affairs and Dispute Resolution in Engineering and Construction. 2018;**10**:04518002. DOI: 10.1061/(ASCE)LA.1943-4170.0000253

[19] Ahmadi M, Behzadian K, Ardeshir A, Kapelan Z. Comprehensive risk management using fuzzy FMEA and MCDA techniques in highway construction projects. Journal of Civil Engineering and Management. 2017;**23**:300-310. DOI: 10.3846/13923730.2015.1068847

[20] Subramanyan H, Sawant PH, Bhatt V. Construction project risk assessment: Development of model based on investigation of opinion of construction project experts from India. Journal of Construction Engineering and Management. 2012;**138**:409-421. DOI: 10.1061/(ASCE)CO.1943-7862.0000435

[21] Li J, Zou PXW. Fuzzy AHP-based risk assessment methodology for PPP projects. Journal of Construction Engineering and Management. 2011;**137**:1205-1209. DOI: 10.1061/(ASCE)CO.1943-7862.0000362

[22] Abdelgawad M, Fayek AR. Risk management in the construction industry using combined fuzzy FMEA and fuzzy AHP. Journal of Construction Engineering and Management. 2010;**136**:1028-1036. DOI: 10.1061/(ASCE)CO.1943-7862.0000210

[23] Zhang G, Zou PX. Fuzzy analytical hierarchy process risk assessment approach for joint venture construction projects in China. Journal of Construction Engineering and Management. 2007;**133**:771-779. DOI: 10.1061/(ASCE)0733-9364(2007)133:10(771)

[24] Rohaninejad M, Bagherpour M. Application of risk analysis within value management: A case study in dam engineering. Journal of Civil Engineering and Management. 2013;**19**:364-374. DOI: 10.3846/13923730.2012.744770

[25] Panthi K, Ahmed SM, Ogunlana SO. Contingency estimation for construction projects through risk analysis. International Journal of Construction Education and Research. 2009;**5**:79-94. DOI: 10.1080/15578770902952181

[26] Dawood N. Estimating project and activity duration: A risk management approach using network analysis. Construction Management and Economics. 1998;**16**:41-48. DOI: 10.1080/014461998372574

[27] Zou PXW, Zhang G, Wang J. Understanding the key risks in construction projects in China. International Journal of Project Management. 2007;**25**:601-614. DOI: 10.1016/j.ijproman.2007.03.001

[28] Zhang Y. Selecting risk response strategies considering project risk interdependence. International Journal of Project Management. 2016;**34**:819-830. DOI: 10.1016/j.ijproman.2016.03.001

[29] Zhang Y, Fan Z-P. An optimization method for selecting project risk response strategies. International Journal of Project Management. 2014;**32**:412-422. DOI: 10.1016/j.ijproman.2013.06.006

[30] Zhi H. Risk management for overseas construction projects. International Journal of Project Management. 1995;**13**:231-237

[31] Mohamed O, Abd-Karim SB, Roslan NH. Risk management: Looming the modus operandi among construction contractors in Malaysia. International Journal of Construction Management. 2015;**15**:82-93. DOI: 10.1080/15623599.2014.967928

[32] Chan DWM, Chan JHL, Ma T. Developing a fuzzy risk assessment model for guaranteed maximum price and target cost contracts in South Australia. Facilities. 2014;**32**:624-646. DOI: 10.1108/F-08-2012-0063

[33] Creedy GD, Skitmore M, Wong JKW. Evaluation of risk factors leading to cost overrun in delivery of highway construction projects. Journal of Construction Engineering and Management. 2010;**136**:528-537. DOI: 10.1061/(ASCE)CO.1943-7862.0000160

[34] Shrestha A, Chan T-K, Aibinu AA, Chen C, Martek I. Risks in PPP water projects in China: Perspectives of local governments. Journal of Construction Engineering and Management. 2017;**143**:05017006. DOI: 10.1061/(ASCE)CO.1943-7862.0001313

[35] Chan APC, Yeung JFY, Yu CCP, Wang SQ, Ke Y. Empirical study of risk assessment and allocation of public-private partnership projects in China. Journal of Management in Engineering. 2011;**27**:136-148. DOI: 10.1061/(ASCE)ME.1943-5479.0000049

[36] Osei-Kyei R, Chan APC. Risk assessment in public-private partnership infrastructure projects: Empirical comparison between Ghana and Hong Kong. Construction Innovation. 2017;**17**:204-223. DOI: 10.1108/CI-08-2016-0043

[37] Li CZ, Shen GQ, Xu X, Xue F, Sommer L, Luo L. Schedule risk modelling in prefabrication housing production. Journal of Cleaner Production. 2017;**153**:692-706. DOI: 10.1016/j.jclepro.2016.11.028

[38] Chan DWM, Chan APC, Lam PTI, Wong JMW. Empirical study of the risks and difficulties in implementing guaranteed maximum price and target cost contracts in construction. Journal of Construction Engineering and Management. 2010;**136**:459-507. DOI: 10.1061/(ASCE)CO.1943-7862.0000153

[39] Jha KN, Devaya MN. Modelling the risks faced by Indian construction companies assessing international projects. Construction Management and Economics. 2008;**26**:337-348. DOI: 10.1080/01446190801953281

[40] Ling FYY, Hoi L. Risks faced by Singapore firms when undertaking construction projects in India. International Journal of Project Management. 2006;**24**:261-270. DOI: 10.1016/j.ijproman.2005.11.003

[41] Wiguna IPA, Scott S. Relating risk to project performance in Indonesian building contracts. Construction Management and Economics. 2006;**24**:1125-1135. DOI: 10.1080/01446190600799760

[42] Santoso DS, Ogunlana SO, Minato T. Perceptions of risk based on level of experience for high-rise building contractors. International Journal of Construction Management. 2003;**3**:49-62. DOI: 10.1080/15623599.2003.10773035

[43] Rostami A, Oduoza CF. Key risks in construction projects in Italy: Contractors' perspective. Engineering, Construction and Architectural Management. 2017;**24**:451-462. DOI: 10.1108/ECAM-09-2015-0142

[44] Han SH, Kim DY, Kim H. Predicting profit performance for selecting candidate international construction

projects. Journal of Construction Engineering and Management. 2007;**133**:425-436. DOI: 10.1061/(ASCE)0733-9364(2007)133:6(425)

[45] Goh CS, Abdul-Rahman H, Samad ZA. Applying risk management workshop for a public construction project: Case study. Journal of Construction Engineering and Management. 2013;**139**:572-580. DOI: 10.1061/(ASCE)CO.1943-7862.0000599

[46] Fernandez-Dengo M, Naderpajouh N, Hastak M. Risk assessment for the housing market in Mexico. Journal of Management in Engineering. 2013;**29**:122-132. DOI: 10.1061/(ASCE)ME.1943-5479.0000128

[47] Adafin J, Rotimi JOB, Wilkinson S. Risk impact assessments in project budget development: Quantity surveyors' perspectives. International Journal of Construction Management. 2018:1-15. DOI: 10.1080/15623599.2018.1462441

[48] Adafin J, Rotimi JOB, Wilkinson S. Risk impact assessments in project budget development: Architects' perspectives. Architectural Engineering and Design Management. 2016;**12**:189-204. DOI: 10.1080/17452007.2016.1152228

[49] Adedokun OA, Ogunsemi DR, Aje IO, Awodele OA, Dairo DO. Evaluation of qualitative risk analysis techniques in selected large construction companies in Nigeria. Journal of Facilities Management. 2013;**11**:123-135. DOI: 10.1108/14725961311314615

[50] Dada JO, Jagboro GO. An evaluation of the impact of risk on project cost overrun in the Nigerian construction industry. Journal of Financial Management of Property and Construction. 2007;**12**:37-44. DOI: 10.1108/13664380780001092

[51] Skorupka D. Identification and initial risk assessment of construction projects in Poland. Journal of Management in Engineering. 2008;**24**:120-127. DOI: 10.1061/(ASCE)0742-597X(2008)24:3(120)

[52] Hwang B-G, Zhao X, Chin EWY. International construction joint ventures between Singapore and developing countries: Risk assessment and allocation preferences. Engineering, Construction and Architectural Management. 2017;**24**:209-228. DOI: 10.1108/ECAM-03-2015-0035

[53] Hwang B-G, Zhao X, Ong SY. Value management in Singaporean building projects: Implementation status, critical success factors, and risk factors. Journal of Management in Engineering. 2015;**31**:04014094. DOI: 10.1061/(ASCE)ME.1943-5479.0000342

[54] de la Cruz MP, del Cano A, de la Cruz E. Downside risks in construction projects developed by the civil service: The case of Spain. Journal of Construction Engineering and Management. 2006;**132**:844-852. DOI: 10.1061/(ASCE)0733-9364(2006)132:8(844)

[55] Perera BAKS, Rameezdeen R, Chileshe N, Hosseini MR. Enhancing the effectiveness of risk management practices in Sri Lankan road construction projects: A Delphi approach. International Journal of Construction Management. 2014;**14**:1-14. DOI: 10.1080/15623599.2013.875271

[56] Chileshe N, Kikwasi GJ. Critical success factors for implementation of risk assessment and management practices within the Tanzanian construction industry. Engineering, Construction and Architectural Management. 2014;**21**:291-319. DOI: 10.1108/ECAM-01-2013-0001

[57] de Marco A, Mangano G. Risk and value in privately financed health care projects. Journal of Construction Engineering and Management.

2013;**139**:918-926. DOI: 10.1061/(ASCE)CO.1943-7862.0000660

[58] Odeyinka HA, Lowe J. An evaluation of risk factors impacting construction cash flow forecast. Journal of Financial Management of Property and Construction. 2008;**13**:5-17. DOI: 10.1108/13664380810882048

[59] Karakhan AA, Gambatese JA. Identification, quantification, and classification of potential safety risk for sustainable construction in the United States. Journal of Construction Engineering and Management. 2017;**143**:04017018. DOI: 10.1061/(ASCE)CO.1943-7862.0001302

[60] Choe S, Leite F. Assessing safety risk among different construction trades: Quantitative approach. Journal of Construction Engineering and Management. 2017;**143**:04016133. DOI: 10.1061/(ASCE)CO.1943-7862.0001237

[61] Nguyen A, Mollik A, Chih Y-Y. Managing critical risks affecting the financial viability of public-private partnership projects: Case study of toll road projects in Vietnam. Journal of Construction Engineering and Management. 2018;**144**:05018014. DOI: 10.1061/(ASCE)CO.1943-7862.0001571

[62] Ling FYY, Hoang VTP. Political, economic, and legal risks faced in international projects: Case study of Vietnam. Journal of Professional Issues in Engineering Education and Practice. 2012;**136**:156-164. DOI: 10.10r61/(ASCE)EI.1943-5541.0000015

[63] Tembo Silungwe CK, Khatleli N. An analysis of the allocation of pertinent risks in the Zambian building sector using Pareto analysis. International Journal of Construction Management. 2018:1-14. DOI: 10.1080/15623599.2018.1484853

[64] Edwards PJ, Bowen PA. Risk and risk management in construction: A review and future directions for research. Engineering, Construction and Architectural Management. 1998;**5**:339-349. DOI: 10.1108/eb021087

[65] Tranfield D, Denyer D, Smart P. Towards a methodology for developing evidence-informed management knowledge by means of systematic review. British Journal of Management. 2003;**14**:207-222. DOI: 10.1111/1467-8551.00375

[66] Betts M, Lansley P. Construction management and economics: A review of the first ten years. Construction Management and Economics. 1993;**11**:221-245. DOI: 10.1080/01446199300000024

[67] Latham M. Constructing the Team: Joint Review of Procurement and Contractual Agreements in the UK Construction Industry. London: Department of the Environment; 1994